D1345859

Post, Mine, Repeat

Helen Kennedy

Post, Mine, Repeat

Social Media Data Mining
Becomes Ordinary

palgrave
macmillan

Helen Kennedy
Department of Sociological Studies
University of Sheffield
Sheffield, UK

ISBN 978-1-137-35397-9   ISBN 978-1-137-35398-6   (eBook)
DOI 10.1057/978-1-137-35398-6

Library of Congress Control Number: 2016938708

This Palgrave Macmillan imprint is published by Springer Nature
The registered company is Macmillan Publishers Ltd. London

ACKNOWLEDGEMENTS

Much of the research discussed in this book received funding from UK Research Councils and from other sources, and for this I'm very grateful. Thanks to funders as follows:

- The Arts and Humanities Research Council (AHRC), which supported 'Understanding Social Media Monitoring' (Grant AH/L003775) with a Research Fellowship. This funding made possible the research discussed in Chapter 6, some of the research discussed in Chapter 8, and the writing of this book.
- The Engineering and Physical Sciences Research Council (EPSRC), which supported 'Digital Data Analysis, public engagement and the social life of methods' (Grant 95575108) and 'Public Engagement and Cultures of Expertise' (Grant EPSRC-CCN2013-LEEDS), discussed in Chapter 4.
- Various small grants from the University of Leeds, including:

  - HEIF V (Higher Education Innovation Fund, Reference 95559240)
  - IGNITE (References 95559274 and 95559274)
  - The Creative and Cultural Industries Exchange (Reference 95559284).

These grants funded interviews in social media data mining companies, discussed in Chapter 5, and focus groups with social media users, discussed in Chapter 7.

Some of the material in this book is derived in part from journal articles, although it has all been cut, edited, re-worked and re-purposed for inclusion here, so it may no longer be recognisable as originating in these spaces. I'm grateful to the publishers of the following journals (and articles) for permission to develop this material here:

- Kennedy, H. (2012) 'Perspectives on sentiment analysis', *Journal of Broadcasting and Electronic Media*, 56(4): 435–450 (Chapter 5), copyright Taylor & Francis/Society, available online at: http://www. tandfonline.com/doi/pdf/10.1080/08838151.2012.732141.
- Kennedy, H., Elgesem, D. and Miguel, C. (2015) 'On fairness: user perspectives on social media data mining', *Convergence: the international journal of research into new media technologies* 21(4): PAGES (Chapter 7), copyright Sage Publications, available online at: doi:10.1177/1354856515592507.
- Kennedy, H., Moss, G., Birchall, C. and Moshonas, S. (2015) 'Balancing the potential and problems of digital data through action research: methodological reflections', *Information Communication and Society* 18(2): 172–186 (Chapter 4), copyright Taylor & Francis/Society, available online at: http://www.tandfonline.com/doi/abs/10.1080/1369118X.2014.946434?journalCode=rics20#.Vel0Y854jGt

My AHRC Fellowship funded research visits to some of the best places in the world for thinking about the issues this book addresses. I'm very grateful to the following people for arranging visits to the following places:

- Principal Researcher Nancy Baym, Microsoft Research Social Media Collective, spring 2015.
- Professor José van Dijck, Comparative Media Studies, University of Amsterdam, summer 2014.
- Professor Helen Nissenbaum, Media, Culture and Communication, New York University, spring 2014.

I'm extremely thankful that so many people, including social media users and employees in social media data mining companies, in public sector and in other organisations, have been willing to spend time talking to me and my researchers about their experiences with mining social data. Without them this book would not have been possible. Another group of people without whom this book wouldn't exist are the many collaborators, post-doctoral research fellows and research assistants with whom I have worked and who have contributed to my thinking and to the words on these pages. These include: Christopher Birchall, Stephen Coleman, Ivar Dusseljee, Dag Elgesem, Julie Firmstone, Jan-Japp Heine, James Hennessey, Patrick McKeown, Cristina Miguel, Stylianos Moshonas, Giles Moss, Matthew Murphy, Katy Parry, Christoph Raetzsch, Stuart Russell, Helen Thornham and Nancy Thumim. Thanks to all of you.

During the time I've been writing this book, there have been many exchanges, presentations, seminars, workshops and conferences around the world which have helped me to develop my ideas. I'm especially grateful to these people for stimulating conversations, online and off: David Beer, David Hesmondhalgh, Noortje Marres, Thomas Poell, José van Dijck and the many brilliant thinkers at the Digital Methods Initiative at the University of Amsterdam and at the Microsoft Research Social Media Collective in Boston, USA. And a big thank you to all the folks at the Universities of Leeds and Sheffield, my former and current employers, two great universities in two great Yorkshire cities, who helped me finish this book in all sorts of ways.

Everybody knows some Daves, and if you're lucky, you know some good ones. I do. David Hesmondhalgh (always a supportive, critical friend) and David Beer took the time to read a draft of this manuscript and to provide really helpful comments, and I'm very grateful to both them for that. My favourite Dave, though, is the one I live with. He drew the picture on the front of this book with a pencil and a piece of paper—no datafication there—and he learnt a bit about data in society in the process. He puts up with me and provides the jokes. Thanks Dave.

Friends and family make writing books bearable, and I'm grateful to every one of them who bothered to say 'how's the book?', especially those who said it repeatedly. This, my friends, is how the book is. The things that interest my kids are wonderfully different from the things that this book is about, and that more than anything keeps me happy and sane, so thank you for being yourselves Pearl and Lonnie. I would wait excitedly with you for the delivery of a packet of loom bands or a special edition videogame book forever. I end with an acknowledgement of the local mums—and some dads—who form invisible networks of support the world over and keep it all going with love and laughs. This is for you (but you don't have to read it).

CONTENTS

LIST OF FIGURES

LIST OF TABLES

CHAPTER 1

Social Media Data Mining Becomes Ordinary

Data Abundance and Its Consequences

In the spring of 2014, a professor heading up a new data analytics centre at a top UK university told an audience gathered to find out more about big data that if all of the data that existed was printed out on A4 paper, it would make a pile so high that it would extend to Jupiter and back seven times. By the time readers encounter these words, the figure will be much higher. Around the same time, the online technology dictionary Webopedia stated that in 2011, we created 1.8 trillion gigabytes of data, enough to fill nearly 60 billion, 32 gigabyte iPads. That's 'enough iPads to build a Great iPad Wall of China twice as tall as the original' (Webopedia 2014). In our big data times, such tales about data quantities abound.[1]

Stories about the qualitative consequences of data quantities also abound, as data get mined, analysed and aggregated by an increasingly diverse range of actors for an equally diverse range of purposes. One widely circulated anecdote tells the tale of the online department store Target and a teenage girl who changed her shopping habits. Where she once bought scented hand cream, she switched to unscented creams. She began to buy dietary supplements—calcium, magnesium and zinc. Target data analysts had previously identified 25 products whose purchase contributed

---

[1] But the data deluge has been with us for some time: as early as 1997, David Shenk wrote that one week-day edition of the *New York Times* contained more data than someone living in the seventeenth century would have encountered in the whole of his or her lifetime (Shenk 1997).

© The Editor(s) (if applicable) and The Author(s) 2016
H. Kennedy, *Post, Mine, Repeat*,
DOI 10.1057/978-1-137-35398-6_1

to a 'pregnancy prediction score' (one journalist described this as Target 'data-mining its way to your womb' (Hill 2012)) and this young woman's score was high. The analysts concluded that she was pregnant and the store started to target pregnancy-related products at the teenager, a move to which her father vehemently objected. But the store was right: the teenager was pregnant, and the store knew this before her family did (Duhigg 2012; Hill 2012).

Another consequence of data abundance, this time on social media, can be seen in the story of a young man who joined a choir when he started college and Facebook's unveiling of his actions. Taylor McCormick joined the Queer Chorus when he started studying at the University of Texas. The president of the chorus added McCormick to the chorus's Facebook group, unaware that McCormick's membership of this group would subsequently become visible to his Facebook friends, including his strict Christian father. A member of a conservative church, McCormick's father did not speak to his son for weeks after the revelation. According to an article in the *Wall Street Journal*, McCormick was the victim of the lack of control we have over our data once they are digitised, or over our lives once they are datafied (Fowler 2012). McCormick 'lost control of his secrets' according to the article.

Digital data mining is used to predict a wide range of phenomena. Increasingly, influence and reputation are matters of numerical prediction through digital reputation measurement platforms like Klout, Kred and Peer Index. These systems produce 'scores' that serve not only as measures of present influence but also as predictive targets of the future. These scores are then used in a number of ways: by hotel chains to determine upgrade rates; by events organisers to give preferential access to parties; in the evaluation of job applications in the digital industries; and by customer services departments to decide how quickly to reply to enquiries— the logic here is that it is better to respond quickly to someone with a high reputation score, as that person will influence more people when talking positively or negatively about his or her experience with a given brand (Gerlitz and Lury 2014). Writing about transactional and financial data, Mark Andrejevic (2013) points to other ways in which data is used to make predictions, including a story from the *New York Times* about credit card companies watching purchasing habits for signs of imminent divorce (Duhigg 2009). Did you use your credit card to pay a marriage counsellor? If you did, your credit might be reduced, because divorce is expensive and affects people's ability to make credit payments. In another example cited by Andrejevic, the Heritage Health Prize set up a competition hosted by Kaggle, a company that specialises in crowdsourcing predictive algorithms, using anonymised health data to produce an algorithm that best predicts which patients might be hospitalised. Such prediction could result in useful, pre-emptive health intervention or, more ominously, in reduced options for health insurance, Andrejevic points out.

In addition to stories about data abundance, their predictive capacities and increasingly disturbing significance, there are also signs of a growth in public awareness of data mining practices. At the time of writing, Take This Lollipop (Zada 2011), an interactive Facebook application, had more than 15.5 million Facebook likes, somewhat ironically, given that Facebook is the target of the app's critical message. On the Take This Lollipop homepage, users are asked—or dared—to take a lollipop by signing in with their Facebook account details. A video runs, in which a menacing-looking man, with dirty, chewed nails, taps on a keyboard and stares at a screen. He's staring at you, looking at your Facebook content, accessed via Facebook's Application Programming Interface (or API), which allows software applications to interact with Facebook data. He's looking at photos of you, or perhaps of your children, tagged as you, and of your friends. He looks at a Google map, identifies where you live, and gets in a car. He's coming after you, getting closer. The video ends with the name of one of your Facebook friends randomly selected by the app's algorithm: 'you're next', you are told about your friend.

A number of similar applications, usually subjecting the user to less visceral experiences than Take This Lollipop, emerged in the 2010s. These testify both to growing awareness of the possibilities and consequences of the mining of social media and other web-based data and to a desire to spread such awareness among web and social media users. They include sites such as We Know What You're Doing … (Hayward 2012) and Please Rob Me (Borsboom et al. 2010), which re-publish public data shared on social media, including views about employers, information about personal alcohol and drug consumption, new phone numbers and absences from home. We Know What You're Doing … describes itself as a 'social networking privacy experiment' designed to highlight the publicness and mine-ability of social media content. If users scroll down to the bottom of the webpage, they discover that the sentence started in the site's title ends with the words '… and we think you should stop'. Similarly, the footer of the Please Rob Me website declares 'our intention is not, and never has been, to have people burgled'. On the contrary, both of these sites aim to raise awareness of the consequences of over-sharing. There are many other examples of this kind.

Another indication of the growth in awareness of digital data mining can be seen in the kinds of articles and reports that populate the pages of the mainstream press with growing regularity. A quick glance at the tabs I have open in my browser on the day I write these words, Thursday, 9 October 2014, demonstrates this. They include: a feature from *The Guardian* newspaper online, from 8 October, entitled 'Sir Tim Berners-Lee speaks out on data ownership' (Hern 2014), with a subtitle which highlights that the inventor of the web believes that data must be owned by their subjects, rather than corporations, advertisers and analysts; a report from *Wired* magazine's website which asserts that our colleagues pose bigger threats to our privacy than hackers (Collins 2014); and another report from that day, this time on the BBC's news webpages, about Facebook vowing to 'aggressively get rid of fake likes' on its platform (BBC 2014).

At the time of writing, anecdotes, apps and articles such as those discussed here—which point to an abundance of digital data, some of the consequences of data mining, and efforts to respond to these phenomena—are ever more common. They attest to the fact that data on our online behaviour are increasingly available and mined, and that the people whose data are mined appear, at least at first glance, increasingly aware of it. They also show that data mining has consequences, which go beyond the outing of young gay people, the withdrawal of credit and the refusal of entry to networking events. As many writers have argued, data mining and analytics are about much more than this: they also offer new and opaque opportunities for discrimination and control. Numerous writers have made this case, including Andrejevic (2013), Beer and Burrows (2013), boyd and Crawford (2012), Gillespie (2014), Hearn (2010), Turow (2012) and van Dijck (2013a), to name only a few. The expansion of data mining practices quite rightly gives rise to criticisms of the possibilities that they open up for regimes of surveillance, privacy invasion, exclusion and inequality, concerns implicit in one way or another in the above examples.

These criticisms, which I discuss in detail in Chapter 3, are entirely justified when it comes to the spectacular forms of data mining and analysis that have hit the headlines in recent years, as carried out by the National Security Agency (NSA) in the US and Government Communications Headquarters (GCHQ) in the UK, as well as governments, law enforcers and the major social media corporations themselves (Lyon 2014; van Dijck 2014). However, at the time of writing, there are many more forms of data mining than these. The expansion of data mining in recent times means that a diverse range of data mining practices exists today, carried out by a variety of actors, in distinct contexts, for distinct purposes, and some of them are more troubling than others. Therefore, we need to differentiate types of data mining, actors engaged in such practices, institutional and organisational contexts in which it takes place, and the range of purposes, intentions and consequences of data mining. Writing specifically about one source of data, social media, José van Dijck and Thomas Poell (2013, p. 11) state that 'all kinds of actors—in education, politics, arts, entertainment, and so forth', as well as police, law enforcers and activists, are increasingly required to act within what they define as 'social media logic'. Such logic, they argue, is constituted by the norms, strategies, mechanisms and economies that underpin the incorporation of social media activities into an ever broader range of fields. One such norm or mechanism is data mining. Given the ubiquity of social media logic, we need to be attentive to the diversity of social media data mining that takes place within the varied fields identified by van Dijck and Poell, in order to fully comprehend data mining in its contemporary formation.

Couldry and Powell (2014) make a similar argument about the need to ground studies of data mining and analytics in real-world, everyday practices and contexts. Acknowledging that enthusiastic belief in the power of data needs to be tempered, they nonetheless argue that:

> However misleading or mythical some narratives around Big Data (…), the actual processes of data-gathering, data-processing and organisational adjustment associated with such narratives are not mythical; they constitute an important, if highly-contested 'fact' with which all social actors must deal. (Couldry and Powell 2014, p. 1)

They argue that the focus in much critical debate on the ability of algorithms to act with agency (they give the work of Scott Lash 2007 as an example) leaves little room to explore the agency of small-scale actors who are making organisational adjustments to accommodate the rise of data's power. In contrast, Couldry and Powell argue that these actors deserve to be examined, alongside 'the variable ways in which power and participation are constructed and enacted' (2014, p. 1) in data mining's practices. In this, they acknowledge that they are echoing Beer's response to Lash, in which he argued that there is a need to focus not only on the power of algorithms, but also on 'those who engage with the software in their everyday lives' (Beer 2009, p. 999). Couldry and Powell propose the same in relation to data mining, arguing that what is needed is an open enquiry into 'what actual social actors, and groups of actors, are doing under these conditions in a variety of places and settings' (2014, p. 2). Evelyn Ruppert and others (2013) make the same argument, stressing the need to focus on the particular ways in which digital technologies get integrated into specific organisations and contexts.[2]

In these debates, a number of terms are used to characterise the types of data mining practices that emerge as social media and big data logics penetrate more and more social spheres. Beer uses the concept 'everyday lives', and Couldry and Powell write about 'smaller' 'social actors', actors with social ends beyond the accumulation of profit, power and control. In this book, I use a similar term to characterise the types of social media data mining that are its focus: ordinary. This term has its roots in cultural studies' insistence on the political gesture of 'lowering' academic sights (McCarthy 2008) to 'activities in the daily round' (Silverstone 1994). At once slippery, contested and fraught, its multiple definitions capture the sense in which I use it here: the commonplace, the apparently mundane, 'the conventional, the normal, the natural, the everyday, the taken for granted' (Silverstone 1994, p. 994), opposite to the extraordinary, the special and the remarkable (McCarthy 2008), 'ordinary daily humdrum' (van Zoonen 2001, p. 673, cited in Thumim 2012, p. 43). Although there is a long history of interest in the ordinary across a broad range of social sciences, what cultural studies did, starting with Raymond Williams's famous 1958 essay 'Culture is ordinary', was to politicise this interest, to ask how and where power operates within ordinary cultures. Taking the phrase from Mukerji and Schudson (1991), in 1994 Silverstone argued that attending to 'the power of the ordinary' signifies a commitment to and a preoccupation with 'the exercise of, and struggles over, institutional and symbolic power' (1994, p. 994) at the level of the everyday. This remains true today, and these questions about the operations of power concern me in relation to data mining, as I explain below.

---

[2] Looking further afield, beyond debates focused specifically on data and digital society, more arguments for the need for such attention to detail can be found. For example, Cruikshank opens the introductory chapter of her book *The Will to Empower: democratic citizens and other subjects* (2000)—the chapter is tellingly called 'Small things'—with a call to arms from Foucault for 'a meticulous observation of detail' (Foucault 1984).

The term 'ordinary' retains its original usefulness in a context of pervasive social media logic, so I use the term 'ordinary' in this book (mostly without inverted commas) in a way that brings together the meanings noted, to refer to the everyday-ness, the commonplace-ness and the ubiquity of social media data mining. Such practices might also be considered mundane, in comparison with the spectacular, high-profile instances of data mining of the kind unveiled by Snowden. Writing about the growth of audit cultures beyond their professional origins (a phenomenon not unrelated to the subject of this book, I will suggest) at the turn of the century, Strathern (2000) highlights why the mundane matters. By themselves, audit practices seem mundane, she states, but when put together with other parts of a bigger picture, they become significant in broader—and troubling—processes of social change. The same is true of data mining practices, as I argue throughout this book.

There is an important difference between my use of the term 'ordinary' and its use by other writers cited here. While they usually write about ordinary *people*, my focus is on *ordinary organisations* doing social media data mining. The data mining that is undertaken by social media platforms, or by large businesses such as the retail companies that Turow et al. (2015a, b) write about, is ordinary in the sense that shopping and being on Facebook are everyday activities for many people. But I use the term 'ordinary' to characterise who is doing the data mining—not governments, security agencies and multinational corporations, but rather local city councils, local museums and other organisations whose operations contribute to structuring daily life. Whether ordinary *people* can mine their own data in the same way that companies and organisations do raises whole set of issues that are beyond the remit of this book. That's another project, and I'll do it next.

To date, most academic attention has focused on the data mining activities of the big and significant actors. These include the mega social media platforms, especially Facebook (for example, in the work of Fuchs, among many others—see Fuchs 2014 and Fuchs and Sandoval 2014), and more recently, in the light of Edward Snowden's revelations, security agencies and national governments (see Lyon 2014; van Dijck 2014). But as I have pointed out, these major players are not alone in submitting to social media/big data/data mining logic. Given that so much attention has already been paid to social media corporations and governmental and security agencies, what we now need to attend to is other, more ordinary actors, as social media data mining becomes ordinary. This, then, is the first objective of this book: to focus attention on the social media data mining activities undertaken by ordinary actors. These include intermediary social media insights companies, which mine and monitor the activity of social media users on behalf of paying clients from public, charity, third and other sectors. They also include public sector organisations, such as local councils, museums and universities, which deploy social media data mining strategies for diverse purposes. Academics across a range of subject specialisms and disciplines are increasingly including social media data mining methods among their toolsets, and social and community groups are also engaging in the mining of digital data.

With this broad range of ordinary social media data mining practices in mind, it is then possible to ask what kinds of social media data mining should concern us, and what should concern us about them. Should we view a resource-poor public sector organisation like a museum or local council, which uses social media data mining in order to understand, engage and provide services for its publics, in the same way that we view the activities of the NSA, or Facebook's failure to distinguish private from public when it trades data assets with third parties? These are the types of questions that the proliferation of social media data mining demands that we address. Then, having assessed what should concern us within this range of practices, the central, normative question of this book can be addressed: can social media (and other) data mining ever be considered acceptable or be used in ways that we find acceptable? Social media data mining raises serious questions in relation to rights, liberties and social justice, and the fact that the 'data delirium' (van Zoonen 2014) is driven by the agendas of big business and big government should trouble those of us who doubt whether these agendas serve the public interest. But it is still important to ask the questions posed here, which draw attention to the current diversity of data mining practices, precisely because more and more actors and organisations are compelled to participate in them and because the uses of data mining have become so varied.

Historically, social studies of technology have asked whether technological ensembles of all kinds can be appropriated as tools of democratisation, inclusion and enablement, despite their origins within the belly of the beast. It seems to me that the same questions should be asked of social media data mining, a new kind of technological ensemble. This concern is captured nicely in the first words of Andrew Feenberg's preface to *Transforming Technology*: 'must human beings submit to the harsh logic of machinery, or can technology be fundamentally redesigned to better serve its creators?' (2002: v). Often what emerges in answer to this question is recognition of the interplay between dominant forces and resistant or appropriating practices which negate the possibility of *complete* domination and control. Applied to the subject matter of this book, this means asking whether there are forms of data mining and analytics that can enable citizens, publics, social groups and communities in the same way that other digital technologies have been seen to do (from video advocacy projects with disenfranchised youth in the 1980s or for human rights purposes (for example Dovey 2006) and the 'vernacular creativity' (Burgess 2006) made possible by online video sharing platforms, to uses of social media in resistance movements (Gerbaudo 2012) and experiments with ICTs for development (Davies and Edwards 2012), to give just a few examples).

So we need to ask: Are there ways in which data mining might make a positive contribution to society? Are there forms of data mining that are compatible with the objectives of small-scale public organisations or community groups, which are likely to be distinct from those of government and corporate actors, or does data mining always have a political bias in preference of these latter groups (Bassett 2015)? These questions take as a starting point the criticisms of data mining referenced above and discussed in detail in Chapter 3. That is to say, because data mining is often used in troubling ways, it is important to consider whether a different and better relationship between it and public life is possible.

This book, then, aims to contribute to debates about social media data mining in the ways indicated here. It focuses on some of the emerging, ordinary practices and actors engaged in it and it assesses these against normative questions about how and whether they should concern us. It addresses a historical question about technology and society with regard to the contemporary example of social media data mining: does social media data mining always inevitably suppress human well-being or are other alternatives possible? If we are committed to a better, fairer social world, we must consider these questions.

In the book, I use the term 'social media data mining' to talk about a broad range of activities which are undertaken in order to analyse, classify and make sense of a social media data. Here, the term 'social media data' refers to what is said and shared, who is saying and sharing it, where they are located, to whom they are linked, how influential and active they are, what their previous activity patterns look like and what this suggests about their likely preferences and future activities. Other terms are sometimes used to refer to these activities, such as 'social media monitoring', 'social media insights', 'social media intelligence', 'social media research' or 'social media analytics'. Sometimes, the words 'media' or 'data' are dropped, and actors talk about 'social data', 'social insights' or 'social media mining'. Some suggest that the term 'social *media* data' refers to the content of posts and tweets and is not as valuable as 'social data', the more significant metadata associated with content (that is, all of the things listed above after the initial item 'what is said and shared'). None of the terms is particularly dominant, although some actors are strong advocates of their own favoured terms. I choose 'social media data mining' because it communicates its referent clearly—that is, the mining on social media of everything that can be considered data. The discussion so far has drawn on debates about data mining more broadly; in the next chapter, I explain why I focus specifically on *social media* data mining, and provide a fuller description of what it entails.

In the years in which I have been researching and writing this book, the term 'big data' has gained and lost traction. Indeed, this chapter starts with an anecdote from a meeting about big data, in which data's bigness was emphasised with an alluring story about piles of data stretching to Jupiter and back. But as quickly as big data rose to glory, it is now diminishing from view, seen as hype and hyperbole, a bandwagon too quickly jumped upon by all kinds of actors. This book is not about big data, because the size of the datasets I discuss is sometimes very small, and because it is not data's *size*, but its *power* that matters in contemporary society. So, this book is better characterised as about data power in society than about big data. As such, it is also about datafication. Mayer-Schönberger and Cukier, who coined this term, describe it like this: 'to datafy a phenomenon is to put it in a quantified format so that it can be tabulated and analysed' (2013, p. 78). The transformation into data of aspects of social life that formerly were not datafied (friendships, relationships, liking things, locations, professional networks, exchanging audio-visual media (van Dijck 2014)) and the related assumptions about what data are and can do, whether big or small, is a fact of contemporary social life. It is in this context that social media data mining has become ordinary. In a sense, then, this is a book about what it feels like to live with datafication and data power, what they constrain, and what they make possible.

The following chapters paint a picture of living with a particular aspect of datafication, social media data mining, as undertaken by actors in ordinary organisations. In subsequent chapters, I argue that, as social media data mining becomes ordinary, as we post, mine and repeat, new data relations[3] emerge, and these are increasingly integral to everyday social relations. These new data relations bring with them a pervasive desire for numbers, I suggest. I characterise this desire for numbers as a convergence of Ted Porter's ideas, developed in the mid-1990s, about the trust that numbers inspire because of their apparent objectivity and facticity (Porter 1995), and Benjamin Grosser's more recent argument that the metrification of social life on social media platforms, produces a 'desire for more' (Grosser 2014). To talk about a desire for numbers, rather than a trust in numbers, makes it possible to account for contradictions that accompany ordinary social media data mining, such as hunger for and evangelism about, but also frustration in and criticism of data and data mining. This widespread desire for numbers brings with it some troubling consequences: it becomes increasingly difficult to discuss problems with social media data mining despite recognition of them, and it has effects of all kinds on work and workers. Despite these problems, and because of the ubiquity of data and data mining that accompanies datafication, the possibility of doing good with data (and with data mining) endures. Together, these and other contradictory tendencies—the persistence of some old concerns, the emergence of new ones, data power and challenges to it—constitute the new data relations that I map out in the pages that follow. But before I do this, I introduce the research on which this book is based and the structure of the book.

RESEARCHING ORDINARY SOCIAL MEDIA DATA MINING

The next two chapters provide some framing for the empirical chapters that follow them. The first, Chapter 2 'Why study social media data mining?', explains why the book focuses on *social media* data mining. I argue that there are good reasons for doing this. First, as we have already seen, social media and their logic are increasingly pervasive. Second, at least theoretically, it is possible for a wider range of actors to mine social media data than other types of data, because the open APIs (Application Programming Interfaces, which allow other software to interact with them) of social media platforms permit this, and because free social

---

[3] Thanks to David Beer for suggesting that this book is mapping a set of 'new data relations'—this is his term, not mine. I take the opportunity to develop it here.

media data mining software makes this possible. But the phrase 'at least theoretically' is an important caveat, and I explain why in later chapters. Third, the use of social media for 'intimacy interactions' (Ito et al. 2009) means that people have particular attachments to what they share on social media, and this is relevant to the question of what should concern us about data mining in these spaces. The chapter then offers a detailed description of what social media data mining is, its tools, processes and applications. The second 'framing' chapter, Chapter 3 'What should concern us about social media data mining?', maps out the concerns that have been raised about contemporary data mining in relation to its more extraordinary forms. These include: the view that the proliferation of data mining and analysis opens up the possibility of new forms of privacy invasion and surveillance; the ways in which data mining is deployed in the interests of discrimination; the black-boxing of the tools of data mining and the subsequent invisibility of the ways in which they make and shape data; and issues of access and exclusion. In the second half of the chapter, I review debates about structure and agency, which I suggest as a framework for imagining whether alternative regimes of data mining are possible.

As this book is about the moment that social media data mining became ordinary, it draws on research undertaken during that moment, the early 2010s, with diverse actors. The research is discussed in Chapters. 4–8, and I describe it here chronologically, as it happened, rather than following the structure of the book. Much of the research has been collaborative, and I acknowledge my collaborators here and elsewhere in the book, aware that their input and conversation have informed my thinking enormously.

I first became interested in social media data mining in 2010 when, at the University of Leeds, my place of work at that time, third year new media undergraduate student James Hennessey suggested that, for his final project, he could develop a sentiment analysis engine. He proceeded to do just that, single-handedly building language libraries, writing algorithms and designing interfaces in a period of just a few months. The resulting application, Social Insight, did a good job of assessing the sentiments expressed on social media (using the limited range of positive, negative and neutral commonly used in this field), until he stopped maintaining it and turned his mind to greater things. It earned him the best mark ever given out for a final year new media project. His work on Social Insight and the problems he faced during its production caused me to reflect on some of the ethical questions raised by mining social media data, so I undertook a series of interviews with workers in social media data mining companies to explore these issues. At the time, some of these called themselves social media monitoring companies, others social media insights companies (alert perhaps to the surveillant connotations of the term 'monitoring') and others provided a broad range of digital marketing services. I undertook 14 interviews between in 2011 and 2013, some of them with the research assistance of Cristina Miguel. The interviews aimed to establish the work biographies of the interviewees, explore the work undertaken by their companies, identify ethical codes or codes of practice to which they adhered, and to put to the interviewees some of the criticisms of social media data mining that were emerging in academic circles, inviting them to reflect on their own ethical positions in relation to these concerns.

The interviews provided insights into what happens inside companies which do social media data mining, but more extensive presence in these contexts was needed in order to reach a deeper understanding of how they work. So I organised internships in digital marketing agencies for two final year undergraduate students in the summer of 2012. Matthew Murphy and Stuart Russell were keen to get experience both of social media marketing and of academic research, which these positions allowed, as they were simultaneously required to participate as interns in work activities and report to me on their reflections on their experiences. These internships were supported by University of Leeds funds which aim to facilitate the development of relationships between academics and the cultural industries, and so they were paid, unlike so many of the virtually compulsory internships that young people undertake today in their efforts to secure paid employment in the future (Perlin 2011). I do not draw on them extensively in this book, but they helped me to paint a picture of the data mining undertaken in digital marketing agencies. The research described in these two paragraphs is discussed in Chapter 5 'Commercial mediations of social media data'.

Around this same time, I worked with colleagues in the former Institute of Communications Studies (ICS), now the School of Media and Communication, at the University of Leeds, to explore the place of data mining in the activities of public sector organisations, as part of a larger initiative focused on the digital transformations that affect communities and cultures, funded by the UK Engineering and Physical Sciences Research Council (EPSRC). Colleagues involved in this study included Stephen Coleman, Julie Firmstone, Katy Parry, Giles Moss, Helen Thornham and Nancy Thumim (see Coleman et al. 2012, for a summary of the research undertaken). I worked closely with Giles Moss to examine how digital data methods (such as statistical analysis, social network analysis, issue network analysis, exploratory content analysis and data visualisation) were being used and might be used by public sector organisations to engage their publics, and to explore what the implications of such methods might be. We carried out interviews within five public sector organisations to explore these questions. As uses of social media data mining methods were emergent in the public sector at the time, we were interested in how digital data methods *could* be used as well as how they were already being used. One way that we explored this was to identify companies offering social media and web analytics services and to carry out textual analysis of their website content, analysing the types of claims they made about their services. We did this with the help of Patrick McKeown and Christopher Birchall. At the end of this brief project, we ran a workshop for the public sector workers whom we had interviewed, which was led by a social media analyst from a commercial insights company which specialised in advising the public sector. This workshop introduced participants to free insights tools which could enable them to advance their existing practices, without having to invest financial resources, which were limited for all of the organisations involved. We ran this event because participants had said that they hoped to gain such knowledge through their collaboration with us. Events like these constitute research methods in that, with participants' consent, they provide data about the perceived possibilities and limitations of tools and techniques, how digital data methods get framed in discourses about them, and how people working in the public sector think about them.

An action research project grew out of this preliminary exploration of the uses of digital methods by public sector organisations. Also funded by the EPSRC, it was called 'Digital Data Analysis, Public Engagement, and the Social Life of Methods'. I worked on this project in 2013 with Giles Moss, Christopher Birchall and Stylianos Moshonas, all colleagues within ICS at Leeds at the time. Working with three of the city-based public sector organisations with whom we had previously carried out interviews, we were interested in exploring ways of circumventing the threat of a new digital divide based on differential levels of data access (boyd and Crawford 2012) discussed in Chapter 3, by examining ways in which resource-poor groups who want to use digital methods for the public good might be able to access them. In the context of the austerity measures of the time, public sector organisations fell into this category: they, like many others, were in danger of ending up on the wrong side of the divide. We also wanted to explore the ways in which 'the public' might come into being through digital methods. Thus we explored the application of digital methods, evaluated their potential use and reflected on their normative consequences. In the prior study, we had identified that our partners were already using some digital methods, but these were not used systematically and, while partners were keen to do more, resources were limited. Thus three of the former collaborating organisations welcomed the opportunity to work with us to experiment with and reflect on the potential usefulness of specific digital methods for their public engagement agendas. As part of this small-scale action research project, I experimented with some of the social media data mining tools discussed in the next chapter. These engagements with the public sector are discussed in Chapter 4 'Public sector experiments with social media data mining'.

From the autumn of 2012 to the spring of 2013, with Dag Elgesem of the University of Bergen and Cristina Miguel of ICS, Leeds, we carried out ten focus groups with 65 social media users to explore their perspectives on social media data mining, with specific reference to the mining of their own social media data. At the time, what users thought about social media data mining had not been addressed directly, although some quantitative studies into attitudes to digital data tracking more generally had been undertaken, and a small number of qualitative studies into attitudes to privacy and surveillance on social media existed. We felt that, as part of a bigger project to understand social media data mining, we needed to ask social media users directly about their knowledge of and attitudes to data mining practices. We also felt that it was important to move beyond dominant privacy/surveillance paradigms, to open up a space for social media users to respond to real-world, ordinary social media data mining practices (described to them in focus groups) in their own words. Social media data mining professionals helped us design this aspect of the research, which was also supported by University of Leeds funds to facilitate relationship-building across academic and cultural industry sectors. The focus groups took place in England, Norway and Spain, the native countries of the three of us, and are discussed in Chapter 7 'Fair game? User perspectives on social media data mining'.

Reading the criticisms of data mining that I discuss in Chapter 3 'What should concern us about social media data mining?' (which characterise it as surveillant, discriminatory and an instrument of control), interviewing social media insights workers, and talking to people in the public sector who were experimenting with data mining, I started to wonder what happened to mined social media data, especially as some people working in commercial companies told me, anecdotally, that the reports they deliver to clients are often simply filed away on virtual desktops or in physical drawers. I wondered if the problematic consequences of data mining were only a problem if data mining results in actual, concrete action and change. So what becomes of the data that are gathered through social media data mining? With whom are they shared within organisations, where do they go, how are they used, if at all, and what impact do they have? And how do all of these things vary across organisations, sectors, practices and actors? These seemed like important questions in the light of criticisms. To explore them, in 2014, with the help of post-doctoral researcher Stylianos Moshonas, I carried out interviews within a range of organisations which do social media data mining, usually by commissioning the services of commercial insights companies. Getting people to agree to be interviewed about this is not easy, because companies and their staff often prefer to keep data mining activities under the radar. Occasionally, though, data mining enthusiasts are keen to talk about the value of the data mining that they do, so I was able to secure interviews in ten companies which, it could be argued, collectively produce the structures of ordinary, everyday life: local authorities, media organisations, higher education institutions, finance-related institutions and cultural organisations. These were supported by an Arts and Humanities Research Council (AHRC) Fellowship and are discussed in Chapter 6 'What happens to mined social media data?'

Across this research, I have carried out interviews, run focus groups, done participant observation and listened to what people say. Throughout the book, I quote participants' words as evidence to support arguments I am building, arguments that did not exist prior to the research—I did not know what I would find when I embarked on it. There is a danger that attending to the experiences of 'what actual social actors, and groups of actors, are doing' (Couldry and Powell 2014, p. 2) in the context of data power can result in an unproblematic acceptance of the words of respondents and a limited acknowledgement of the broader political economic context in which data mining takes place. But while my approach could be described as phenomenological, in the sense that it attends to the points of view of actors and their experiences, this is not because I overlook the conditioning contexts which structure how data mining takes place. Rather, I give priority to empirical detail because in discussions of data power to date, the emphasis has been on structuring forces.[4] The empirical detail in this book is intended to complement critical commentary on structures of data power, not to dismiss or overlook it. This is a long-winded way of saying that there is a lot of empirical description in this book and some people might not like it, but it is here for good reason.

In addition to the empirical research detailed above, over the past few years, I (and researchers working with me) have attended a number of social media data mining events, some academic, some commercial, and some aiming to bring these two sectors together. I refer to these events occasionally in the subsequent chapters. Academic events have exposed me to issues and approaches in academic social media data mining. The intentions, processes and outcomes of academic research which makes use of these methods are usually much more public than those of corporate and commercial actors anyway, so by attending academic conferences, reading journals and participating in mailing lists and other online discussions, it is possible to develop some understanding of uses of social media data mining for academic social research without having to study it in the same way as other sites. I discuss academic social media data mining in Chapter 8 'Doing good with data: alternative practices, elephants in rooms', alongside discussion of forms of data activism, which I also have not studied empirically. Data activism includes open data initiatives, citizen science, campaigns for better and fairer legislation in relation to data, and movements which seek to evade dataveillance like Anonymous. While not strictly focused on *social media* data (although some are), these activities seek to imagine better ways of living with data mining. I put academic social media data mining and data activism together because they are both underpinned, to different degrees, by a desire to do good with data, as the chapter title suggests, and this is important in relation to the issue of what should concern us about social media data mining. But before moving on to any of these data mining sites and practices, I start with some framing.

---

[4] This is changing. At the time of writing this book, I organised a two-day, international conference called Data Power, which included many papers based on empirical studies of data's power in specific context. The conference programme can be found here: http://www.sheffield.ac.uk/socstudies/datapower/programme.

CHAPTER 2

Why Study Social Media Data Mining?

INTRODUCTION

This chapter does two things. First, it provides a rationale for the book's focus on social media data mining. Then it describes in detail what social media data mining is. As the examples at the start of the previous chapter show, the expansion of data mining stretches beyond social media platforms, but in this book I focus primarily on the mining of *social media* data. I argue that there are good reasons for doing so, as the characteristics of social media make the data that are generated, mined and analysed within them an important object of study.

The first important characteristic of *social media* data mining is that it is possible for a wide range of actors to do it. This is because the open Application Programming Interfaces (APIs) of social media platforms, mentioned in the previous chapter, allow third party software applications to access their data, making it possible for 'ordinary' actors to interact with and analyse them. This is not the case for other types of data—transactional data, financial data, health-related data, for example, are much less easy for ordinary actors to access, as they are usually proprietary and private, or available at a high cost from information and data trading companies.[1] Building on these open APIs, a number of free social media intelligence tools are available, some of which are discussed later in the

[1] Of course, there are important questions about whether any of these data should be available for mining, as subsequent chapters indicate.

© The Editor(s) (if applicable) and The Author(s) 2016   19
H. Kennedy, *Post, Mine, Repeat*,
DOI 10.1057/978-1-137-35398-6_2

chapter, making it widely possible to gain insights into social media activity. Thus the mining of social media data is more open and accessible than other forms of data mining, at least in theory—this is an important caveat, which I return to later in the book.

Second, as I discuss below, social media platforms are often used for what Ito et al. (2009) call 'intimacy interactions'. Social media are perceived by users as spaces for sharing feelings and thoughts, for building and maintaining relationships, for forming and performing identities, and for doing these things frequently and informally. As such, they are spaces in which the personal and intimate aspects of our lives are lived out; Trottier (2012) describes them as places of 'interiority'. Consequently, we might expect people's attachments to their social media data to be different from their attachments to other data generated in less intimate circumstances. Scrutinising what happens to social media data is important, given these attachments. Finally, as suggested in the previous chapter, the 'logic' of social media is increasingly pervasive (van Dijck and Poell 2013). Social media platforms penetrate many aspects of everyday life, shaping and reshaping people's personal interactions and organisational operations. As such, the mining of data produced on social media platforms, which constitutes one element of their extended logic, demands attention.

In the next section, I say more about the distinctive characteristics of social media, starting with some definitions and statistics about usage patterns. I argue that the distinctive qualities of social media data make its mining an object worthy of scholarly attention. To focus on *social media* data mining alone is to separate social media out from the ecosystem of data mining and aggregation in which it exists, but nonetheless, I argue that this is a valuable exercise, because of the characteristics of social media which are fleshed out in this chapter. These include: their participatory character; the invocation to share on social media; the various ways in which they can be considered intimate; and their monetisation. Then, having presented a rationale for my focus on social media data mining, I provide a sketch of approaches and applications. I classify available tools into four types: in-platform, free-and-easy, commercial, and free-and-complex. In a field in which new tools and technologies surface regularly, as I observe below, by the time this book is published, some of the technologies I discuss will have faded into history and others will have emerged as major players in the social insights landscape. But perhaps the categorisations I offer to guide the non-expert reader through this terrain will endure.

FOUR CHARACTERISTICS OF SOCIAL MEDIA: PARTICIPATION, SHARING, INTIMACY AND MONETISATION

Initial efforts to characterise the new 'social' media that emerged as a result of the growth of Web 2.0 technologies can be traced to boyd and Ellison's widely cited definition of social networking sites, or SNSs, in the mid-2000s. In this definition, the authors asserted that SNSs were typified by users' ability to:

(1) construct a public or semi-public profile within a bounded system, (2) articulate a list of other users with whom they share a connection, and (3) view and traverse their list of connections and those made by others within the system. (boyd and Ellison 2007, p. 211).

In a later publication, in 2013, Ellison and boyd note that SNSs have evolved dramatically since that early definition: some features have decreased in significance, some have been adopted by other online genres, and there has been a general merging of SNSs and other platforms through the connecting and exchanging of data with open APIs. Van Dijck (2013a) draws on a definition of social media from Kaplan and Haenlein (201) to articulate the relationship between SNSs and social media, the term I use in this book. According to Kaplan and Haenlein, social media are 'a group of Internet-based applications that build on the ideological and techno-logical foundations of Web 2.0, and that allow the creation and exchange of user-generated content' (2010, p. 60; cited in van Dijck 2013a, p. 4). Within this definition, argues van Dijck, these four types of social media can be identified:

- **social networking sites** (SNSs), which, as boyd and Ellison argue, promote interpersonal contact and communication, examples of which include Facebook, Twitter, LinkedIn and Google+;
- **user-generated content** (UGC) sites, which promote the exchange of creative content produced by amateurs and professionals (and indeed arguably blur the distinction between these groups), such as YouTube, Flickr, Instagram and Wikipedia;
- **trading and marketing sites** (TMSs) for exchanging or selling products, such as Amazon, eBay or Craigslist;
- **play and game sites** (PGSs) such as FarmVille, The Sims and Angry Birds.

But as van Dijck points out, social media platforms experiment across these classifications and merge elements of categories in efforts to dominate their fields. Facebook, for example, encourages users to share content just like UGC platforms, as well as experimenting in both trading/marketing and play/games.

Statistics attest to the unquestionable popularity of social media. In May 2013, Facebook reported that 4.75 billion pieces of content were shared daily (Facebook 2013). At the same time, there was reportedly an average of 400 million tweets posted every day (Twitter 2013). In early 2014, there were over 1.15 billion Facebook users, 1 billion enabled Google+ accounts (albeit compulsory to set up when other Google-related accounts are established), 550 million registered users on Twitter, and two image-sharing sites, Pinterest and Instagram, had over 20 million and 150 million active monthly users respectively. At that time, 72% of all internet users claimed to be active on social media, with that figure rising to 89% among 18–29-year-olds, and still as high as 60% among 50–60-year-olds (Bullas 2014). A statistical bulletin on internet access produced by the Office for National Statistics (ONS) in the UK reported that, in 2013, 73% of adults in Great Britain accessed the internet everyday (ONS 2013), with that figure rising as high as 93% for 16–24-year-olds. And alongside these dominant platforms (which also include LinkedIn, YouTube and Flickr), many less well-known platforms exist.

The growth in use of social media platforms has been accompanied by a parallel growth in academic studies of them. These studies emerge across academic disciplines and, as such, address a broad range of issues, from psychological questions about what motivates people to self-disclose openly on social media, to socio-cultural questions about the impact of widespread social media use on, for example, community, identity and political participation, and legal questions about the governance of these platforms. Among these debates, the opportunities that social media open up for participation (also referred to as collaboration, produsage, citizen media-making and more), have received widespread attention. As Kaplan and Haenlein suggest through their definition of social media as allowing 'the creation and exchange of user-generated content' (2010, p. 60), these basic activities of users creating, exchanging, sharing and collaborating have led social media, the foundational Web 2.0 platforms, to be understood to enable a fundamentally participatory culture.

But early celebrations of the democratic and inclusive potential for Web 2.0 participation, for example in the work of Benkler (2006), Bruns

(2008) and Jenkins (2006), have been quickly tempered with more cautionary analyses which question assumptions about the participatory character of so-called participatory networked cultures. In 2009, van Dijck highlighted the low numbers of active participants in participatory cultures with reference to the (scientifically unproven) '1% rule': 'if you get a group of 100 people online then one will create content, 10 will "interact" with it (commenting or offering improvements) and the other 89 will just view it' (2009, p. 44). Van Dijck argued that it is therefore important 'to distinguish different levels of participation in order to get a more nuanced idea of what participation entails' (2009, p. 44). A special issue of *Communication Management Quarterly* published in 2011 aimed to do just that, critically interrogating the notion of participation and asking whether users 'really participate' (Carpentier and Dahlgren 2011). Especially relevant to the aims of this book is the recognition that participation 'has the potential to cut both ways', as Andrejevic puts it, making possible both 'the increasing influence of participatory consumers on the production process, and the facilitation of monitoring-based regimes of control' (2011, p. 612). In other words, participation produces the very data that gets analysed in social media data mining processes.

While academic analyses focus on the concept of participation in social media, this term is not widely used by the platforms themselves. Instead, what the platforms invite us to do is to 'share'. Nicholas John (2013) provides a comprehensive account of the history of social media platforms' invocation to share, tracing the use of the term across 44 major SNSs from 1999 to the time of publication. He identifies 2005–2007 as a time when platforms started to use the language of sharing. Bebo, for example, moved from inviting users to 'Write and Draw on other peoples' (sic) White Boards' and 'Keep in contact with friends at other Universities' to describing itself thus: 'Bebo is a social media network where friends share their lives and explore great entertainment' and encouraging users to 'Invite Friends to Share the Experience'. Similarly, Fotolog moved from encouraging users to 'Make it easy for friends/family to see what's up with you' to inviting them to 'Share your world with the world' (cited in John 2013, p. 175). By the mid-2010s, the ubiquity of share buttons on social media platforms and other online spaces testified to the centrality of the concept for the platforms themselves.

But, as John points out, despite its widespread use on social media platforms, and despite the comprehensive study of other terms used to refer to similar things, the concept of sharing has been surprisingly under-studied (although Wittel (2011) and Belk (2010) are exceptions). In his historical analysis, John notes that whereas users were once invited to share specific objects, such as photos on Flickr, in more contemporary usage of the term, the invitation is to share either what he calls 'fuzzy objects', such as our lives or our worlds, or no objects at all, through calls to 'connect and share!' or simply 'share!' Thus sharing has become both more vague and more inclusive. This use of the term, argues John, assumes we know what is meant by it—we know what it is that we are being invited to share; we do not need this to be specified. At the same time, such usage is dense, because so many activities fall under the classification 'sharing', such as status updating, photo sharing, book reviewing, tweeting, commenting and communicating. Thus sharing describes in new ways activities which were not previously defined in this way.

John suggests there are a number of reasons why sharing has come to be so widely used on social media platforms: the term was readily available because of its historical use in computing (for example file sharing, disk sharing and time sharing); it is versatile in its reference both to the communication of feelings and the distribution of digital content; and it has connotations of equality, selflessness and of positive social relations. There are other important reasons for its use too, which are relevant to the focus of this book, as the language of sharing serves to obscure the commercial character of these large-scale, multinational corporations, enabling them to represent themselves as fundamentally desiring a better world. Platforms promote the notion that the more we share, the better the world will be, as seen in the letter Mark Zuckerberg attached to Facebook's 2012 initial public offering (IPO) filing:

> people sharing more—even if just with their close friends and families—creates a more open culture and leads to a better understanding of the lives and perspectives of others. (cited in John 2013, p. 177)

Sharing is also the term of choice for many social media platforms to describe their commercial transactions with third parties, precisely because of the connotations discussed above, which 'paper over the commercial aspects of the ways in which many SNSs operate' (2013, p. 177), as John puts it. Both Google and Facebook, for example, talk about 'sharing personal information' in their terms and conditions and privacy policies.

So far I have suggested that social media platforms can be thought of as spaces for participation and for sharing. They can also be considered as spaces for 'intimacy interactions' (Ito et al. 2009). For despite heated debate about whether people care about privacy in the age of social media, fuelled in part by Facebook CEO Mark Zuckerberg's assertion in 2010 that 'privacy is no longer a social norm', what is shared, and what we are invited to share, is material that in other contexts would be considered private, personal and intimate. This is evidenced, for example, in Facebook's invitation to users to post on their timelines the answer to this personal question: 'What's on your mind?' As De Ridder (2015) argues, social media platforms promote the practice of being social, thus transforming intimacy from a private to a public matter, though many have contested the value of this simple private/public distinction in the context of social media, as I highlight in the next chapter. Trottier (2012) concurs with this view of social media as spaces to be intimate. Tracing the history of social media across various platforms, he argues that what they have in common is that they are characterised by 'interiority' (2012, p. 6). They are perceived as spaces for psychological comfort, in which communication is phatic, not informational. But as well as being spaces in which personal information is shared, social media can be said to be intimate in other ways too, suggests Trottier. First, he argues that social media are dwellings; we spend a lot of time in them and use them extensively. We depend on dwellings for privacy, and the walls of dwellings are supposed to shield us from public scrutiny. But on Facebook, walls become spaces for the public display of our personal lives. For him, one of the problems with social media data mining is that it is comparable to surveilling our private dwellings.

Another way in which social media activity might be described as intimate is that platforms share many of the characteristics of ubiquitous computing—they are perpetually operational, and so pervasive they become imperceptible to the human eye. I take this to be one of the multiple meanings of Kember and Zylinska's concept of 'life after new media' in their book of that name (2012). Social and other new media go unseen by many people as they go about their everyday lives, but it is this very invisibility of social media *as* media, or as technology, that gives the data produced in social media a more intimate status than, say, financial data, which is often more knowingly and consciously shared. The material devices of social media matter in this regard too. With most social media activity taking place via mobile technologies (53% in the UK in 2013, according to the ONS (2013)), this device + platform combination achieves the ultimate cyborg dream. When the cyborg metaphor was first developed by Donna Haraway (1985) and others in the mid-1980s to articulate the growing intimacy between human and machine, the bodily proximity of technology and flesh was captured through the idea that technologies increasingly 'stick to the skin'. The mobile devices of social media communication do just that—or, if they do not 'stick to the skin', they are at least are very close to it, and this too speaks to the intimate place of social media in users' lives.

Thus social media activity is intimate in a number of ways, as sites like Facebook become extensions of our social lives. For many people, social activity *is* social media activity, because of the presence of many known peers on social media platforms and their widespread use to organise social activity. Social media are spaces for sharing the personal and intimate minutiae of our daily lives, otherwise transient personal activity which is subsequently archived for years to come by platforms. And they are spaces of participation—in communicative, community-forming activities, in the exchange of created and curated cultural goods, and in the forming, through this participation, of social relationships. Trepte and Reinecke argue that on social media, 'People create online spaces of social and psychological privacy that may be an illusion; however these spaces seem to be experienced as private and the technical architecture of the Social Web supports this notion' (2011, p. 62). In other words, when active on social media, people may not feel that they are sharing data with platforms; rather, they feel that they are sharing intimate things, or what John calls fuzzy objects, with intimate others. Data do not have to be shared to be taken, and this is another distinguishing feature of social media. I share my bank account details with an online store when I purchase something, but I do not necessarily feel that I am sharing data about my music tastes with Facebook when I post a photo of a concert I have attended. Facebook, in contrast, considers that I have shared this data with it, and proceeds to 'share' it with paying others.

Whatever the original aims and motivations of social media platforms in encouraging users to share matter of all kinds, today, their content and data have become monetisable assets. This should not be surprising, as Web 2.0 was always intended as a business manifesto—O'Reilly's early maxims included, among others, 'data is the next Intel inside', 'users add value' and 'collaboration as data collection' (O'Reilly 2005, cited in Lister et al. 2009, p. 204). Van Dijck argues that, despite their origins as informal spaces for communication and creativity, today social media platforms 'programme with a specific objective' (2013, p. 6) the activities that take place within them. Thus social media sites that started as community platforms have become global information and data mining sources. On these platforms, relationships and connections are coded into algorithms, and the distinction between human connectedness and automated connectivity is deliberately blurred; connectedness is used to obscure connectivity, argues van Dijck. Facebook founder Mark Zuckerberg's emphasis on making the web social is a coded way of talking about 'making sociality technical' (2013, p. 12). Relationships are commoditised, connectedness is converted into connectivity through code, and both are mined, analysed and sold for profit. Enter social media data mining.

To consider *only* social media data mining (as opposed to a wider range of data and data mining types) is to separate out a process which, in reality, is part of a larger ecosystem (to use van Dijck's term) of digital data tracking, aggregation and analytics, as personal data are increasingly combined and shared across digital spaces. Social media platforms recognise this, although their acknowledgement of it is often distributed across various pages buried deep within platforms' architecture and not easy to find. For example, in information about Facebook advertisements, the company advises:

> Facebook shares the data it gathers about you with third parties. Facebook works with others to combine information they have collected from you with information from elsewhere, to enable the advertiser to send you targeted advertising on Facebook. (Facebook nd-b)

Other examples of data aggregation ecosystems can be found in Turow's book, *The Daily You* (2012), where Turow charts the complex interrelationships between behavioural advertisers, data traders and other relevant actors like information vendors Experian and Acxiom, companies dealing in personal data for targeted advertising, and analytics companies like Audience Science, which track and analyse browsing patterns. He cites a report on the 'Data-driven Web' from 2009, which claimed that:

> contact information is now collected at virtually every step in a user's online experience (via registration pages, for example) and Web surfing behaviour is tracked down to the millisecond—providing publishers and advertisers with the potential to create a reasonably complete profile of their audiences, and thus enabling robust, segmentation-based targeting. (cited in Turow 2012, p. 79)

Not only are 'wide-ranging data points indicating the social backgrounds, locations, activities, and social relationships of hundreds of millions of individuals' becoming 'fundamental coins of exchange in the online world' (Turow 2012, p. 89), but data collected online is linked to data obtained from other sources. Social media data mining takes place within this wider data aggregation landscape.

But the interconnections between ecosystems of data and data mining notwithstanding, social media platforms are distinctive because of their near ubiquitous penetration of what van Dijck and Poell (2013) describe as 'the mechanics of everyday life'. Van Dijck and Poell point to the pervasiveness of 'social media logic', something that the statistics cited earlier confirm. They argue that the four grounding elements of social media logic are: programmability (the ability of platforms and users to steer and influence communication and information exchange); popularity (measured, influenced and manipulated by social media platforms); connectivity (both the enabling of human connectedness and the pushing of automated connectivity); and datafication (the rendering into data matter which formerly was not data, as noted above (Mayer-Schoenberger and Cukier 2013)). It is in relation to this latter element that social media data mining plays an important role.

As van Dijck and Poell see it, the logic of social media extends beyond social media themselves, to invade all areas of public life—not only the media, but also law and order, social activism, politics, education and other spheres. Social media steer how users interact with each other, in personal interactions and institutional and professional contexts, and communication becomes entangled with social media's commercial mechanisms. Van Dijck and Poell suggest that:

> in contemporary society, no institution can afford to look away from this logic because they have all become implicated in the same media culture: every major institution is part and parcel of this transformation in which the social gets infiltrated by a revamped social media logic. (2013, p. 11)

For this reason too it is important to focus on the operations of social media, despite the interconnectedness of mined data. So in this book, I temporarily extract social media from their ecosystem to interrogate the mining that takes place within them by some of the ordinary actors noted by van Dijck and Poell. Chapters 4–8 of this book undertake that task. But first, in the next section, I provide a description of what social media data mining is, how it works and who does it. I finish by introducing some of the problems and issues it raises, which are then taken up in further detail in the following chapter.

What Is Social Media Data Mining?

At a very basic level, social media data mining can mean counting the likes and shares of content that take place on social media. Platforms themselves offer such measurement services, for example in weekly page updates distributed via Facebook to anyone with administrative rights on a Facebook page. New page likes, weekly total reach and other figures are compared with the same statistics for the previous week, to give an indication of whether social media exposure is increasing or decreasing. The other major social media platforms offer similar analyses. Twitter's analytics service (analytics.twitter.com) tells you how many impressions your tweets have earned on 1 day (that is, the number of times users saw your tweet), in comparison with previous days, as well as providing information about engagement (the number of times users have interacted with a tweet) and engagement rates (the number of engagements divided by the number of impressions). It provides information about followers as well, including interests, location, gender, and the people they follow. YouTube users can also access statistics about views, traffic sources and user demographics, and Flickr Pro offers a statistics feature showing view counts and referrer data, which indicate where images have been used on the web.

A growing number of online applications offer similar services to those provided by the platforms, often combining data from across social media into a single dashboard. To give one example, in the museums sector, the website Museum Analytics (http://www.museum-analytics.org/) provides weekly reports on total Facebook page likes, new page likes, posts and comments, as well as total Twitter followers, new followers, tweets and mentions, while also comparing these figures with those of previous weeks, and illustrating statistics with simple graphs such as those reproduced in Figure 2.1. All of these figures are made publicly available on the Museum Analytics website, on a platform-by-platform basis, as well as for individual museums. Museum Analytics compiles qualitative as well as quantitative data: for each museum, the platform shows the five most engaging content items on Twitter and Facebook, measured by retweets on the former, and likes and comments on the latter.

These examples represent the very tip of the iceberg of social media data mining. Online tools like Social Mention (http://socialmention. com/) and TweetReach (http://tweetreach.com/), to give two examples of freely available tools at the time of writing, offer more analysis (and

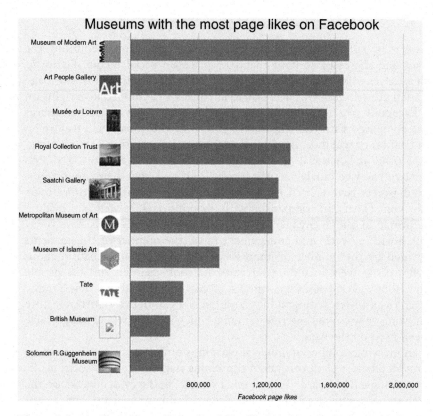

Figure 2.1   An illustration of mined social media data from the Museum Analytics website (source: http://www.museum-analytics.org/facebook/, accessed: 14 October 2014)

more again if you pay a fee for their professional versions). A search for a phrase or keyword on Social Mention will report these things about it:

- strength, given as a percentage, and described as the likelihood of a social media mention, calculated on the basis of total mentions within the last 24 hours divided by total possible mentions;
- sentiment, given as a ratio, and described as the ratio of mentions that are designated positive to those that are designated negative (thus ignoring those that are designated neutral);

- passion, given as a percentage: the likelihood that people mentioning the keyword will do so repeatedly, so a small group of people talking consistently about your topic will result in a higher passion score than a large group doing so occasionally;
- reach: the number of unique authors of mentions divided by the total number of mentions, so lots of authors compared to not many mentions results in a high reach figure;
- how many minutes lapse between mentions;
- when the last mention occurred;
- number of unique authors of mentions;
- numbers of retweets of original tweets;
- top keywords used in mentions;
- top users mentioning the keyword or phrase;
- top hashtags used in mentions;
- top sources (that is, the platforms from which mentions have been identified).

A Social Mention report will also contain all mentions of the searched keyword or phrase, and allows the user to filter mentions by source type—blog, microblog, images, video, comments and so on.

TweetReach, in contrast, focuses solely on the Twitter platform, as the name suggests. Users enter a query, such as a URL, hashtag or phrase, and the platform analyses tweets that it considers match the search. It reports on estimated reach (accounts reached), exposure (number of impressions), activity (such as tweets to retweets ratio, and most retweeted tweets), and contributors (including top contributors). If the system identifies more than 50 relevant tweets, it reports on the first 50, and offers the user the option of buying a report based on the full dataset for $20 at the time of writing. Other platforms and applications offer similar and more advanced options, some for free and others at a cost, and new tools and techniques regularly become available. Those mentioned here are examples of a genre, not the most important, nor significant in themselves. Without doubt, by the time this book is published, some of those discussed here will be defunct, and other, newer tools will have evolved.

The two categories of social media data mining discussed so far might be defined as 'in-platform' and 'free-and-simple' respectively. I now turn to two other categories, which I call 'commercial' and 'free-and-complex'. As with all categories, an ever-changing range of commercial services are available for paying clients. According to one industry insider, Francesco

D'Orazio, at the end of 2013, there were more than 480 available platforms (D'Orazio 2013). It is hard to know which platforms are included
in this figure as D'Orazio does not specify, but other reviews attempt to
name important players in the field (Social Media Biz 2011; Social Media
Today 2013)—Sysomos; Radian6 (subsequently part of SalesForce Cloud);
Alterian SM2; Lithium and Attensity360, most of which are US-based,
and Brandwatch in the UK often figure in these lists. Alongside specialist
social media insights companies, digital marketing agencies increasingly
offer social media intelligence services. These include many of the services
listed above, such as identifying: how many people are talking about a
particular topic or brand across social media; key influencers; where key
conversations are taking place; 'share of voice' or how much conversation
is dedicated to a product or brand; strength of feelings about particular topics and brands; and the demographics of people engaged in social
media conversations, such as geographical location, gender, income and
interests. These services involve monitoring not only social media content
and its movement within platforms, but also the metadata that lies behind
those more visible traces of social media activity. For some, this is more
valuable, and marks a distinction between social *media* data and social
data, as suggested in the previous chapter.

Like the free-and-simple tools discussed above, social media monitoring companies mine the core social media platforms, most notably
Twitter, but also public content on Facebook, YouTube, Instagram,
Tumblr, blogs, forums, and conversations from elsewhere on the web
designated 'social' because they are user-generated, such as the comments
sections of newspapers' websites. Companies use a range of technologies
and methods to deliver their services. Boolean search, which combines
keywords with operators such as AND, NOT and OR, is commonly used.
Some tools map networks, either of connected people or connected content. Methods like sentiment analysis use complex systems such as Natural
Language Processing to instruct computers to make sense of word use,
word combination and word order, and so assign sentiments to social
media content.

Such services come at a high cost. At the time of writing, Brandwatch
licenses its platform for a minimum of £500 ($800) a month (for a lim-
ited number of mentions, as well as limited support, storage and historical
data) and a maximum of £2000 ($3200) a month, for more of every-
thing. Meltwater Buzz, an offshoot of the media monitoring company
Meltwater, charges around £7000 ($11,350) per year, plus tax, for use

of its platform with customer support. Sysomos has two platforms, MAP, available for £20,800 ($30,000) per licence per year, and Heartbeat, costing £9000 ($13,000) per licence per year. Some companies use tools like Brandwatch to produce insights for clients, analysing mined data so that clients don't have to. Costs for such services vary greatly. For example, on a modest, small-scale research project that I undertook, £5000 ($8000) bought us the manual analysis of 400 of the 200,000 tweets on a given topic by a social insights company, which, at the same time, were providing insights in the form of multiple reports to a large international sporting event for a total cost of around £65,000.

In this landscape of costly commercial platforms, a number of developers, often working in academic contexts, have produced social media monitoring tools that are freely available and are often much more transparent in how they work than their commercial counterparts. But while free, these applications bring other costs, notably the time and effort required to learn to use them. As a result, they are sometimes seen as difficult to use by people who do not consider themselves to be experts or technologists, as I show in Chapter 4. Hence I call these 'free-and-complex' tools. These include, for example, NodeXL and Gephi, and some of the tools produced by the Digital Methods Initiative (DMI) in Amsterdam, such as the IssueCrawler. I discuss each of these briefly below.

NodeXL and its associated Social Media Importer plugin (http://nodexl.codeplex.com/) are freely available downloads from the Social Media Research Foundation. They add network analysis and social media connectivity capabilities to Microsoft Excel, running on the Windows operating system only (see Figure 2.2 for an example of a NodeXL interface). NodeXL can harvest data from a variety of sources (Twitter, YouTube, Flickr, Facebook and www hyperlinks), but it can only access one of these sources at a time. Some technical knowledge is needed just to find the application, as it is an Excel template file, not a standalone programme. Once opened, the user needs to understand the tool's terminology in order to make productive use of it. For example, users need to know that choosing 'levels to include' means selecting whether to build networks of followers only (level 1), or followers of followers (level 2), and so on, and it is helpful if they also know that level 1 searches rarely reveal anything interesting. Then, users need to know how to make sense of search results, how to manipulate results so they are visualised in meaningful ways, and how to export results into other tools, like Gephi, to produce visualisations of found data.

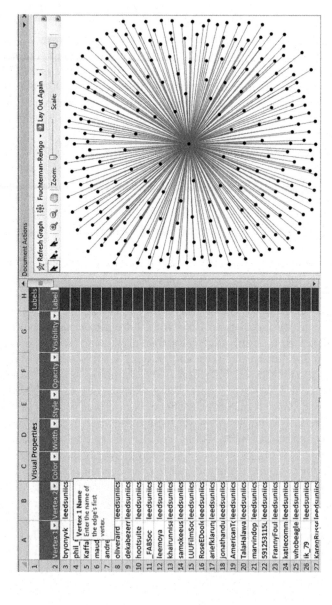

Figure 2.2 The results of a search as displayed in NodeXL

Gephi (https://gephi.org/) is an open source tool for network visualisation—that is, the production of visualisations which allow for analysis of networks and related data. Initially produced by students at the University of Technology of Compiègne (UTC) in France, it is currently supported by a not-for-profit foundation. It allows graphical representation of the 'connectedness' and 'influence' of individuals within a network (it can do a lot more, but these are the most common uses). As Gephi does not have plug-ins that harvest data directly from social networks, datasets need to be produced elsewhere and then imported into Gephi. Figure 2.3 gives an example of a Gephi visualisation.

IssueCrawler is an academic tool for discovering linked content on the web. Users can sign up for free but accounts must be approved by

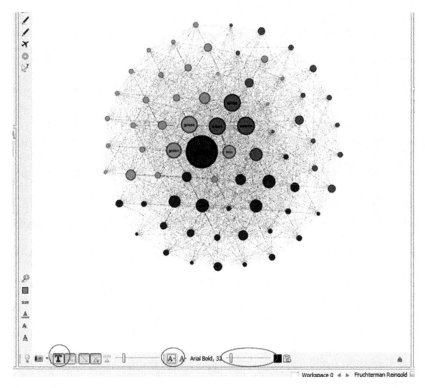

Figure 2.3 Example of a Gephi visualisation

administrators. Like some other tools, IssueCrawler relies upon some prior knowledge of the search topic in order to ensure results. Searches need to be 'seeded' with related URLs, known websites that have content that is relevant to a required search, identified by prior manual investigation such as simple web searches. IssueCrawler uses co-word analysis to build networks of linked content, as it seeks to identify the co-occurrence of search terms across online spaces.

These applications and others with origins in academic contexts are often designed with specific purposes in mind, to meet the aims of the specific projects for which they were developed. New tools are developed with great frequency in academic contexts, as existing tools are often deemed to be not quite fit-for-purpose. As a result, a large number of similar-but-different tools exist, which are difficult to map. Sometimes, companies do the same, developing in-house bespoke tools which are often hard to access and know, because of their originators' interest in keeping their proprietary technology just that, proprietary.

Thus social media data mining tools and techniques come in a range of types, which I have called here in-platform, free-and-simple, commercial, and free-and-complex. So far in the chapter, my discussion of them has been descriptive, as my aim has been to paint a picture of what social media data mining involves. But the description belies a wide range of issues and problems that many writers have raised, which I discuss briefly here and in more detail in the next chapter and later in the book. Thoughtful readers will already be asking themselves: how can something as complex and ephemeral as a sentiment be measured, and then classified according to only three simple categories (positive, negative and neutral)? How do irony, sarcasm, humour and abbreviation affect the results that these tools produce? When searching on the basis of keywords or hashtags, what is missed? What is omitted when these methods are used? Who is not on social media, and therefore not represented in the results of such searches? Which platforms are not searched? How do platforms and applications calculate their results and how are data made and shaped in these processes? What do social media data mean to the people who produce it and what does that mean for what social media data *are*?

Some of these questions are viewed as pragmatic concerns by social media data miners. Sentiment analysts, for example, worry about the effect of sarcasm and irony on their ability to assign sentiment accurately to content, and 70% accuracy is considered good—that is, 70% agreement with what a human would judge a sentiment to be, not 70% accurate identification of sentiments. Less accuracy is common: one social insights expert I interviewed stated that 70% inaccuracy would be a more realistic figure. The disclaimer in the footer of Twitter's analytics website gives another indication of the unlikelihood of accuracy in social media analytics: 'The data reported on this page is an estimate, and should not be considered official for billing purposes' (Twitter Analytics nd). Data miners and analysts are also concerned about the question of what social media data are. For them, what matters is not the philosophical question of how to conceptualise social media data, but rather their reliability. Fake data is an issue for practitioners, either in the form of fake reviews or what Fowler and de Avila (2009) call a 'positivism problem' in online ranking and rating—the fact that the average rating for things reviewed online is 4.3 out of 5 (in the US, it is even higher, at 4.4).

But many of these questions are more political than pragmatic. Social media insights companies want their results to be as comprehensive as possible, as colleagues and I discovered when working with one on a research project, during which they willingly added local forums to their pool of searched platforms, on our suggestion, in order to widen their searches. At the same time, the fact that data are omitted from searches points to the unrepresentativeness of datasets (which are also unrepresentative because whole populations are not active on social media) and to issues of inequality and exclusion. And these questions are also epistemological, as they relate to how data, and therefore knowledge, are made. Lisa Gitelman (2013), among others, highlights these epistemological issues, contesting the possibility of 'raw data' and the term's underlying assumptions of neutrality and objectivity in a book she edited which uses Bowker's (2005) assertion that *'Raw Data' is an Oxymoron* as its title. Social media data, like other data, are not a window onto the world, but are shaped by decisions made about how to go about seeking and gathering those data.

In addition to these political and epistemological questions, social media data mining raises issues which are much more overtly about power—questions about how forms of data mining are used to manage populations, to segregate and discriminate, and to govern and control. Clearly, then, describing social media data mining is not a simple matter. Alert to these questions, which I address in detail in the next chapter, I have nevertheless attempted to describe it here, because it is important to understand what social media data mining is, precisely in order to consider these issues and questions.

CONCLUSION

This chapter has outlined why studying social media data mining is important. Some of the fundamental characteristics of social media platforms make social media data mining a significant object of study, to be extracted briefly from its place in the ecosystem of aggregated and connected data practices. The first is that social media platforms are participatory, but what matters is the potential of participation to 'cut both ways' (Andrejevic 2011, p. 612)—not only to allow users to influence production, but also to open up opportunities for monitoring and monetisation. This, the monetisation of social media data, is another characteristic which is important in understanding why social media matter. Another is that social media platforms invoke sharing. Again, what matters is how this concept cuts multiple ways: it refers to the communication of feelings and the distribution of digital content, and it has connotations of equality and of positive social relations. Because of these characteristics, it is the term of choice of many social media platforms when they describe the commercial sale of user data to third parties. Thus the participation and sharing that take place on social media and are actively encouraged by the platforms allow us to understand both front-end user practices and the more opaque, behind-the-scenes data mining operations. A final important characteristic of social media is that they are spaces for intimacy interactions. As such, people may have particular attachments to their social media data, so scrutinising what happens to social media data is an important endeavour.

So, on the one hand, understanding social media as spaces for relationship building and sustenance, through practices of participation, creative exchange, sharing and intimacy interactions, helps us understand the particular place of social media in social life and the reasons why we need to study the data mining that takes place there. On the other hand, the pervasiveness of social media and their logic also makes social media data mining worthy of academic scrutiny (van Dijck and Poell 2013). The datafication of things that were not previously data and the openness of social media APIs, which allows access to datafied matter, not only make social media logic pervasive but also make social media data mining a real possibility for many more ordinary actors than can access other forms of data and engage in their mining.

In the second half of this chapter, I described some of the processes and technologies that constitute social media data mining. There are many social media data mining applications that I do not mention here. A brief glance at one of the papers referenced in the last chapter gives an indication of some of them. In their paper on digital reputation tools, Gerlitz and Lury (2014) mention these examples, in addition to some of the tools that I discuss: Tweetstats; TwentyFeet; Twittercounter; Crowdbooster; Twitalyzer; Archivedbook; Likejournal; Friendwheel; Touchgraph; Vasande Visualiser; Mentionmap; Paper.li; Keepstream; Twylah; Trunk.ly; Storify. Tools proliferate. The DMI in Amsterdam offers a broad toolset, not just the IssueCrawler, including 58 named applications at the time of writing, and frequently cooperates with the MediaLab at Sciences-Po in Paris to produce even more. At the time of writing, commercial and academic social media data miners alike are addressing the challenge of developing mechanisms for mining the 75% of social data which is untagged, unstructured images, and no doubt some progress will have been made by the time the book is published. Capturing the full range of tools, whether in-platform, free-and-simple, free-and-complex, or commercial, is an impossible endeavour, and I have not attempted to do so. Nor am I an advocate for the tools I have discussed, and I am sure Gerlitz and Lury are not either. Instead, I have discussed a small sample of the tools and processes which actors in the ordinary organisations which are the book's focus might use. I discuss them and their uses of these tools later in the book. First, in the next chapter, I outline how the question of what should concern us about data mining has been addressed in relation to its extraordinary and spectacular manifestations and I consider what other frameworks might be helpful for thinking about more ordinary forms of data mining.

CHAPTER 3

What Should Concern Us About Social Media Data Mining? Key Debates

INTRODUCTION

As more social activities take place online and the social media platforms on which this activity takes place continue to grow, more data on users and their online behaviour becomes available. The growing availability of social data has led to a flurry of bold epistemological claims about what their analysis might tell us. It can be used, it is suggested, to provide new insights into social networks and relationships, analyse public opinion in real time and on a large scale, and capture people's actual behaviour as well as their stated attitudes. The social media data mining methods discussed in the previous chapter can be mobilised, so the narrative goes, to produce knowledge about a range of social phenomena.

Zealous belief in the explanatory potential of data is captured in the widely cited words of *Wired* magazine editor-in-chief Chris Anderson, who proposed that 'every theory of human behaviour, from linguistics to sociology, [...] taxonomy, ontology and psychology' can be dismissed because, in what he calls the Petabyte Age, 'the numbers speak for themselves' (Anderson 2008, np). Anderson's words reflect a belief that data have the potential to transform all aspects of society, making all of its operations more efficient, without the need for analysis, interpretation or theorisation. But despite Anderson's dismissal of the contribution that academics might make to data-driven cultures, they are not immune from the big/social data buzz. In one example, Thelwall et al. state that sentiment analysis, one form of social media data mining, gives researchers the ability 'to

© The Editor(s) (if applicable) and The Author(s) 2016   41
H. Kennedy, *Post, Mine, Repeat*,
DOI 10.1057/978-1-137-35398-6_3

automatically measure online emotion' (2011, p. 408). Nor are such beliefs exclusive to the technical sciences; within the social sciences, faith in the possibilities opened up by big data methods can also be traced. The Oxford Internet Institute's project 'Accessing and Using Big Data to Advance Social Scientific Knowledge' proposes that big data 'represents an opportunity for the social sciences to advance the understanding of human behaviour using massive amounts of data' (Oxford Internet Institute nd). Big data, the project website claims, can be used in the social sciences to 'create new knowledge about the social world and the behaviour of human beings' (Oxford Internet Institute nd). In many of these examples, the focus is big data, defined as characterised by their volume, velocity and variety, or as too big for a desktop computer to process—a somewhat outdated definition, given the capacity of desktop computers. But the adjective 'big' is irrelevant, as I suggested in the introduction to this book, not simply because small quantities of data also matter to ordinary organisations, but because what matters is datafication (Mayer-Schoenberger and Cukier 2013) and the related assumptions about what data are and can do, whether they are big or small. Thus this chapter and others in this book engage with debates about big data, where relevant, even though the data that are my focus are sometimes small, and always social.[1]

Alongside fervent claims like Anderson's, tempering concerns have been raised about 'the data delirium' (van Zoonen 2014), as already noted in previous chapters. Many writers have argued that the proliferation of data, their mining and analysis open up the possibility of new forms of discrimination, exclusion, privacy invasion, surveillance, control, monetisation and exploitation. This chapter focuses on these concerns. As this book asks what kinds of social media data mining should cause us concern and what should concern us about them, I map out the key concerns that have been expressed about data mining to date, in order to consider whether these apply to the ordinary data mining practices that I discuss in subsequent chapters.

Here, I organise criticisms of social media and other contemporary forms of data mining into four categories. The first of these can be broadly characterised as highlighting concerns about privacy and surveillance. Although these two issues are sometimes pitted against each other by writers who argue that a focus on the privacy of individuals fails to acknowledge the more important structural issue of increased social surveillance,

---

[1] See Coté (2014) and Manovich (2011) for more on definitions of social data.

I consider their interrelationship here because, together, they have dominated critical debate about data mining to date. The second area of critique focuses on the ways in which data mining discriminates. This leads to the third set of criticisms, referenced briefly in the last chapter, which relate to methodological concerns about the ways in which data are made and shaped, and the effects this then has on knowledge and decision-making. The final area of criticism addresses the question of who gets to do data mining and issues of access and exclusion: as some writers have pointed out, while we all create data, few of us can access it and even fewer can process it, and this results in the danger of new, data-driven digital divides (boyd and Crawford 2012).

I map out these criticisms in the pages that follow. Collectively, they represent the main concerns that have been raised about data mining in times of datafication. Of course, other schema and categorisations are possible; what I propose here represents just one way of understanding and classifying debate. More criticisms exist than those covered in this chapter, and in the chapters that follow, I discuss some of them in relation to specific data mining practices. In my view, the criticisms discussed here are valid in relation to some of the more spectacular and visible forms of data mining that have hit the headlines, and some are better formulated than others. But I argue that, to date, the focus on structures of datafication has meant that consideration of the possibility of acting with agency against data power has been relatively absent. The concept of agency has been core to many efforts to explore how cultures and societies are made, and how they might be made fairer and more equal. So thinking about agency is one way of addressing the question of whether it is possible to do (social media) data mining in ways that makes a positive contribution to social life. Thus in the second part of this chapter, I highlight ways in which we might think about agency in relation to data mining.

In this chapter, I hope to persuade readers that I take the criticisms I discuss here as a starting point from which to ask: how can small-scale, ordinary organisations and the people within them act in relation to social media data mining? This means thinking about what kinds of actions are possible, as well as whether there is room for agency in data structures. These are ethical questions, so I end the chapter with a discussion of how we might move beyond the notion of agency to think about acting ethically in times of data mining. This, I suggest, is one way of addressing the book's central normative question of whether social media data mining can ever be undertaken in ways that are considered ethically acceptable, and that make a positive contribution to social life.

CRITIQUES OF (SOCIAL MEDIA) DATA MINING

Less Privacy, More Surveillance

In 2010, Facebook founder and CEO Mark Zuckerberg announced that in the age of social media, privacy is no longer 'a social norm' (Johnson 2010).[2] Despite criticism of Zuckerberg's words, there is widespread acceptance of this view—exemplified, for instance, in a *New York* magazine feature in 2007 which dubbed social-media-savvy young people the 'say everything' generation. This generation, the article claimed, think of themselves as having an audience, archive the moments of their adolescence openly, and have a thicker skin than previous generations. A number of commentators have contested this view, such as danah boyd, whose extensive ethnographic research into teen attitudes to social media is reported in her book *It's Complicated: the social lives of networked teens* (2014). Here, she argues that privacy still matters to young people today. Earlier, in a speech in 2010, boyd used the notion of 'being private in public' to highlight the complex relationship between publicness and privacy in social media. Comparing social media with corridors, she argued that if two people bump into each other in the public space of the corridor, one might say something private to the other that s/he would not want to have publicised (boyd 2010). Building on this metaphor, in her book, boyd makes a distinction between being public and being *in* public. Young people might be doing the latter, but they still want privacy, she argues (boyd 2014). This might be understood as 'privacy in context' or what media philosopher Helen Nissenbaum (2009) has termed 'contextual integrity'. Nissenbaum argues that when people disclose personal information in a particular context, they have expectations of what will happen to their data in that context. Ignoring those expectations is a violation of people's data rights. What is needed, then, is contextual integrity, or respecting people's expectations of what might reasonably happen to their data within a given context. Contrary to Zuckerberg's assertion, these arguments suggest that privacy still is, in fact, a social norm.

Of course, it is in the interests of social media companies who make money by selling the matter that users share on social media platforms to

---

[2] Some argue that Zuckerberg did this in order to justify his company's controversial decision to change the default privacy settings of its users' accounts and make them more public than they had previously been.

tell us that we no longer care about privacy—indeed, such strategies play a role in shaping how we think. Yet despite the efforts of Zuckerberg and others to dismiss the significance of privacy in social media environments, the concept retains traction, so that examples of social media privacy 'invasion' such as those cited in this book's introduction are greeted by public concern, and academic researchers seek to understand what they call the 'privacy paradox', or the fact that social media users' sharing practices appear to contradict their expressed privacy concerns. In relation to this 'paradox', some authors conclude that, for users, there is a distinction between social privacy (controlling which people within their networks get access to their information) and institutional privacy (the mining of personal information by social media platforms and other commercial companies) and that social media users' concerns about controlling their personal information relate to the former, not the latter (Raynes-Goldie 2010; Young and Quan-Haase 2013). boyd (2014) provides one possible explanation for this: she argues that we all seek privacy from those who hold power over us and, for many young social media users, the people with power over them are within their own social worlds.

So for some writers, one problem with social media data mining is that it invades personal privacy. Others have suggested that this focus on personal privacy invasion fails to acknowledge the more significant structural issue of the increased surveillance that widespread social media data mining brings. Netichailova (2012), for example, argues that a focus on users' responsibilities to take more care in the management of their personal privacy ignores what she calls 'societal aspects', by which she means the political economic structuring forces within which data mining takes place. Like the threat to privacy, the threat of more surveillance is a major criticism of social media data mining, as high-profile examples of social media surveillance circulate, the Snowden revelations being the most visible and spectacular. Thus concerns about less privacy and more surveillance are sometimes pitted against each other, yet at the same time they have much in common, not least their ability to catch the public eye.

In *Social Media as Surveillance* (2012), Trottier argues that social media data mining is a new form of surveillance (see also Fuchs 2014). Others have argued that alongside institutional and state surveillance, as exemplified in data mining practices undertaken by governments and commercial actors, new forms of social surveillance emerge in social media. This is the phrase that Marwick (2012) uses to describe the reciprocal observation that takes place on social media platforms. Likewise, writing before her,

Albrechtslund (2008) describes online social networking as participatory surveillance; such practices are also described as lateral surveillance or self-surveillance. Humphreys (2011) asserts that participants in his study of the check-in app Dodgeball care about these social and lateral forms of surveillance, not institutional surveillance, echoing the distinction made in privacy-related studies discussed above.

Mark Andrejevic has written extensively about how new digital technologies, including social media platforms, extend surveillance practices (2004, 2007, 2011, 2013). As early as 2004, he argued that interactive technologies allow previously unmonitored activities to be subject to surveillance and thus our digital communication practices become a part of the surveillance society. More recently, writing about a range of practices which include forms of social media data mining like sentiment analysis and opinion mining, but also body language analysis, neuromarketing and drone technology, Andrejevic maps out the various ways in which our actions are subjected to surveillant scrutiny, providing a convincing array of examples to back up his point (Andrejevic 2013). Writing with Kelly Gates about drone technologies' ability to capture all wireless data traffic in an area, he argues that in the big data age, definitions of surveillance change. Formerly understood as 'purposeful, routine, systematic and focused' (Murakami Wood et al. 2006, p. 4, cited in Andrejevic and Gates 2014, p. 189), it is now much less so. Now, drones capture all data about everyone, and store it forever, and surveillance is much more speculative—data is captured for retrospective sorting, on the off-chance that it may be relevant and of use in the future. This is evident in the words of the CIA's Chief Technology Officer:

> The value of any piece of information is only known when you can connect it with something else that arrives at a future point in time. [...] Since you can't connect dots you don't have, it drives us into a mode of, we fundamentally try to collect everything and hang on to it forever. (Sledge 2013, cited in Andrejevic and Gates 2014, p. 185).

Andrejevic and Gates note that there are serious infrastructural challenges for those who would gather all of the data all of the time, which relate to physical storage space as well as tools and techniques. Social media platforms like Facebook have the infrastructure in place that make it possible for communications platforms to serve as surveillance systems; in a recent online list of the ten largest databases in the world, a number of social media platforms figure (Anonyzious 2012, cited in Andrejevic and Gates 2014, p. 189). Thus social media are imbricated in contemporary forms of surveillance, as actors in surveillant practices, as infrastructures that enable surveillance and as databases which house the datasets that are surveilled. Some of the practices revealed by Snowden, and written about by Andrejevic and others, should concern us, as they raise serious questions in relation to human rights, liberties and social justice. But it seems to me that there are two questions we should ask in relation to the concerns I have sketched out in this section.

The first question is whether such concerns apply equally to contemporary forms of data mining, on social media platforms and elsewhere, given the current diversity of data mining practices. This question is addressed in subsequent chapters. The second question relates to what is obscured from view when high-profile examples of privacy invasion and totalising surveillance so easily capture the public imagination. Stories like the Queer Chorus and Target cases discussed in the introduction of this book receive attention precisely because they highlight invasions of personal privacy and the spread of surveillance. The predominance of such stories might be seen as a 'post-Snowden effect', as Snowden's revelations may have created a space in which it is possible to acknowledge how data mining contributes to increased surveillance and privacy violation. But as these stories occupy the limelight, what other aspects of social media data mining are not so visible? One answer to this question relates to methodological issues about how data mining sorts, discriminates and comes to structure social life. The next two sections focus on these matters.

Discrimination and Control

A number of researchers have criticised the discriminatory consequences of data mining. For example Turow, mentioned in the last chapter, has carried out numerous studies in the digital advertising industries in order to unveil how they use data to discriminate. In *The Daily You: how the new advertising industry is defining your identity and your worth* (2012), Turow argues that the data mining that these industries undertake leads to social discrimination. This happens because, through data analytics processes, 'individual profiles' are turned into 'individual evaluations' (2012, p. 6). Calculations of each person's marketing value are produced, based on behavioural and other forms of data tracking, and each individual is categorised as target or waste. These data define our identity and our worth, argues Turow, determining not only what marketing firms do but also how we see ourselves and others. Those of us who are considered to be waste receive narrowed options in the advertising messages that are targeted at us, and, according to Turow, these impact upon our sense of self.

Beer and Burrows (2013) point to the ways in which data discrimination operates in other aspects of culture. Writing about the ways in which music consumption technologies generate archivable data about consumption habits (what kinds of music are consumed, how frequently and for how long, the geographical locations of consumption practices), they map out how these data play a role in constituting the hardware, software and commerce of music consumption in different ways. They argue that data about listening practices feed into 'the production of large-scale national geodemographic systems that in turn provide postcode-level analysis of people's tastes and preferences' (2013, p. 59). As such, data constitutes much more than culture, serving also to shape regimes of governance and control. Elsewhere, writing specifically about sentiment analysis, Hearn argues that data mining's ability to identify valuable sentiments represents yet another capitalist mechanism of value extraction. She describes the people doing this work as 'feeling-intermediaries', who 'structure feelings into profits for themselves and their clients' (2010, pp. 435–43). Likewise Andrejevic (2011) is concerned about the role played by social media sentiment in the prediction and control of affect, which he describes, quoting Massumi, as 'an intrinsic variable of the late capitalist system, as infrastructural as a factory' (Massumi 2002, p. 45, quoted in Andrejevic 2011, p. 609).[3]

These writers highlight the problematic ways in which data mining's discriminatory capacities are put to work, through surveillance, social sorting and other forms of control. But whereas much of this work focuses on *intentional* discrimination in data mining, Barocas and Selbst (2014) draw our attention to the ways that data mining discriminates unintentionally. Data mining, argue Barocas and Selbst, is discriminatory by design. 'The very point of data mining', they write, 'is to provide a rational basis upon which to distinguish between individuals' (2014, p. 6). In a paper which considers whether US law is up to the job of legislating for the discrimination that occurs in the context of data mining, these authors map out the ways in which the distinct technical components of data mining can lead to discrimination. These include: the process of coming up with definitions;

---

[3] Christian Fuchs is another writer who has highlighted how the discriminatory potential of data mining is captured in the interests of capital (for example, Fuchs 2011).

the selection of training data; feature selection; the use of proxies; and masking, or the ability of decision-makers with prejudicial views to make intentional discrimination look unintentional. Thus they draw attention to problems with how data mining works. Through this highlighting of such unintentional processes, Barocas and Selbst point to methodological and epistemological issues that have concerned other commentators on data mining methods. I discuss these writers and their concerns in the next section.

Methodological Concerns

Social media data mining is a method for trying to understand social patterns and, as such, it raises methodological concerns. These concerns are captured to some degree in the idea that methods of all kinds, not just social media data mining, have a social life (see for example Law et al. 2011). Methods are of the social world; they come into being with purposes and advocates. The sample survey, write Law et al., gained momentum in the UK in the 1960s as a means for researching society which bypassed the views of the elite and, as such, was supported—or advocated—by the government of the day. But methods are not only 'shaped *by* the social world in which they are located' they also 'in turn help to *shape* that social world' (Law et al. 2011, p. 2). Methods simultaneously represent aspects of the social world and have social effects, constituting the things they claim to represent (Law and Urry 2004). Like all other research methods, data mining practices are 'performative': 'they have effects; they make differences; they enact realities; and they can help to bring into being what they also discover' (Law and Urry 2004, pp. 392–3). Tarleton Gillespie makes the same point about the algorithms that underlie data mining techniques, describing them as 'socially constructed and institutionally managed mechanisms for assuring public acumen: a new knowledge logic' (2014, np). So it is important to think critically about methods, 'about what it is that methods are doing, and the status of the data that they're making' (Law et al. 2011, p. 7). This means problematising claims, such as those mentioned in the introduction to this chapter, about social media data mining's capacity to provide access to people's opinions, attitudes and behaviour, and instead seeing the data that are produced through these methods as made and shaped.

Human decisions influence and shape the design, development, arrangement and implementation of technological systems and artifacts, as

Science and Technology Studies (or STS) scholars have long argued, and the assemblage of social media data mining is no exception. Gitelman and Jackson (2013) make this point in their introduction to Gitelman's edited collection *'Raw Data' is an Oxymoron* (2013), in which they counter the assumptions implicit in the term 'raw data' (of neutrality and objectivity) with reference to Bowker's claim, paraphrased in the book's title, that 'raw data is both an oxymoron and a bad idea; to the contrary, data should be cooked with care' (Bowker 2005, p. 184). Data do not simply exist, argue Gitelman and Jackson (and so do many others, such as boyd and Crawford 2012). Rather, they need to be generated and, in order to be generated, they need to be imagined. Epistemological claims about what big data analytics can achieve represent one form of imagining. Bollier (2010) adds to this, arguing that to produce data necessarily involves interpretation, and interpretation, in turn, is necessarily biased by the subjective filters that the individual human actors involved in their generation apply, as in the processes described by Barocas and Selbst (2014).

How widely the 'cooking' of data is acknowledged is the subject of some debate. Boellstorff (2013) argues that those working with big or social data understand their datasets as limited representations of the world, conditioned by the theories that framed their generation. Similarly, Havalais (2013) proposes that 'No one expects the massive amounts of data collected by the Large Hadron Collider or various shared astronomical instruments to be free from the influence of the process by which they were collected.' Likewise Barnes notes that the assumption that any scientific knowledge, including that derived from big, social data, is the product of a 'disembodied universal logic' (2013, p. 298) has long been contested. These writers suggest that the limitations of digital data and related methods are widely understood. On the other hand, some critics argue that more awareness is needed of the social shaping of data through the methods that are used to generate them, because there is a 'widespread belief that large data sets offer a higher form of intelligence and knowledge that can generate insights that were previously impossible, with the aura of truth, objectivity, and accuracy' (boyd and Crawford 2012, p. 663). In the introduction to a special issue of the journal *First Monday* on big data, Helles and Bruhn Jensen (2013) concur. They argue that the assumption that data are 'available and plentiful' exists among researchers as well as in public debate, and yet 'the process of generating the materials that come to function as data often remains opaque and certainly under-documented in the published research' (2013).

But although debates about the shaping and making of data through data mining are taking place within academic fields like the social sciences, which are accustomed to subjecting their methods to critical scrutiny, the same cannot be said for actors in small-scale organisations beginning to think about how social media data mining might help them meet their objectives, as the subsequent chapters show. There are many reasons for this. One is because the 'black-boxing' (Pinch 1992) of data mining methods and algorithmic processes makes it difficult to know—and show—how tools and processes make and shape data. The concept of the 'black box', long used in STS, highlights the fact that there is little understanding of the internal workings of the technologies that constitute our social world. Gillespie (2014) argues that this is a problem for various forms of social media data mining, including Twitter's Trend algorithm and digital reputation measurement platforms like Klout, as the criteria by which such systems operate are typically unknown (Gillespie 2014). Another reason that the ways in which data mining construct data may not be transparent to 'ordinary' actors is that data are regularly detached from their conditions of production once 'in the wild'. The Target example, discussed in the book's introduction, is a case in point. Although the company's data analysts might have been aware that they only identified a *likelihood* of the teenage girl being pregnant, once their data became 'numbers in the wild', what were once possibilities became something more concrete. Talking about the movement of data across space, Gerlitz (2015) quotes Espeland and Sauder (2007) who write: 'numbers are easy to dislodge from local contexts and reinsert in more remote contexts. Because numbers decontextualise so thoroughly, they invite recontextualisation.' Finally, in the chapters that follow, I develop the argument that social media data mining is motivated by, produces and reproduces a 'desire for numbers' which has various troubling consequences, including making it difficult to create a space in which to reflect critically on data mining methods. This, then, is another possible reason that, in the public domain, the spotlight is rarely shone on how data are made and shaped. Rather, their apparent objectivity is taken-for-granted. For now, I turn to a final theme in criticisms of data mining: issues of non-representativeness, exclusion and limited access to the skills, knowledge and expertise needed to engage in social media data mining and open up its black box.

Issues of Access and Inequality

Writing in 2012 about growing interest in big data in the social sciences, boyd and Crawford insist on the need to address a number of critical questions about the big data phenomenon, including who gets access to it, and the skills, knowledge and expertise needed to engage in data mining. Thus one of their 'six provocations for big data', as their article is called, is that 'limited access to big data creates new digital divides' (boyd and Crawford 2012, p. 673). They point out that, while much of the enthusiasm surrounding big data comes from a belief that it is easy and straightforward to access, this is not the case. Lev Manovich concurs, stating that 'only social media companies have access to really large social data. [...] An anthropologist working for Facebook or a sociologist working for Google will have access to data that the rest of the scholarly community will not' (Manovich 2011, p. 5). Without doubt, elite commercial companies like Google, Facebook and Amazon have the best access to data, as well as the best tools and methods to make sense of it (Williamson 2014); some companies have even gone as far as discouraging academics from researching social media. Some companies restrict access to data entirely, others sell access for a high fee, some offer small datasets to university-based researchers. So those with money or inside a company have differential access to social media data to those without the necessary financial resources to acquire it. boyd and Crawford (2012) argue that, given this trend, soon only top-tier, well-resourced universities will be able to negotiate access to commercial data sources, and students from those elite, top-tier universities are likely to be the ones invited to work inside social media companies, exacerbating existing patterns of inequality.

Manovich states that there are three ways of relating to big data: there are 'those who create data (both consciously and by leaving digital footprints), those who have the means to collect it, and those who have expertise to analyze it' (2011, p. 10). In the case of data mining, as boyd and Crawford point out, who is deemed to have expertise determines who controls the process and the 'knowledge' about the social world that results, knowledge which in turn reproduces the social world, as we saw in the previous section. 'Wrangling APIs (Application Programming Interfaces), scraping, and analyzing big swathes of data is a skill set generally restricted to those with a computational background' (2012, p. 674), they argue. Manovich confirms this, stating that in 2011, 'all truly large-scale research projects, which use the data with (social media) APIs so far have been done by researchers in computer science' (2011, p. 5). He gives some examples of why this is the case: for instance, if you want to mine the digitised books in Google's Ngram Viewer to analyse changing frequencies in book topics over time, you need expertise in computational linguistics and text mining to do so, he writes.

So access to data, data mining and the skills to do it is uneven and this leads to new digital divides. Thus boyd and Crawford warn that there is a danger of a big data rich and a big data poor emerging in relation to data access, tools, skills and expertise. But why does such access matter? They quote Derrida in order to answer to this question. Derrida claims that an essential criterion by which 'effective democratisation' can be measured is 'the participation in and access to the archive, its constitution, and its interpretation' (Derrida 1996, p. 4, cited in boyd and Crawford 2012, p. xx). Differential access to data archives, how they are constituted and used represent 'inequalities written into the system' (boyd and Crawford 2012, p. xx).

Concerns about data divides do not question the underlying operations of data. Rather, they draw attention to the uneven distribution of data mining and attendant skills, and seek to highlight the problematically undemocratic character of such inequalities. The other criticisms of the mining of social and other data discussed in this chapter focus more explicitly on the discriminatory consequences of data mining and, as such, they point to some of its troubling effects. Criticisms focusing specifically on social media serve to counterbalance the more celebratory accounts of the democratic possibilities opened up by social media and their acclaimed possibilities for participation (for example in Jenkins 2008; see Andrejevic 2011 for a response, cited in the previous chapter). They do the important job of making visible what is largely invisible—that is, the work of monitoring, mining and tracking the data that we leave behind as we move through social media. But they do not address the question of whether human beings must submit to the harsh logics of social media and datafication, or whether it is possible for data mining to be compatible with the interests of small-scale actors, citizens and publics. This question compels us to think about agency, and I turn to this concept in the next section.

Seeking Agency in Data Mining Structures

Baack (2015) argues that thinking about agency is fundamental to thinking about the distribution of data power and yet, in the context of datafication, questions about agency have been 'obscured by unnecessarily generalised readings' (Couldry and Powell 2014, p. 1) of the supposed power of technological assemblages like data mining. For this reason, Couldry and Powell call for more attention to agency than theories of algorithmic power, or data power, have thus far made possible. One way of doing this, they suggest, is through 'social analytics',[4] which they define as the 'sociological treatment of how analytics get used by a range of social actors', or how social actors use data to 'meet their own ends' (Couldry 2014, p. 892). So, as well as acknowledging the problems that the rapid spread of data mining brings with it, we also need to explore the possible agency of actors in ordinary, small-scale organisations in relation to data mining, argue Couldry and Powell (2014). Later chapters in this book could be described as adopting this social analytics paradigm, in their attention to the adjustments made by such actors in response to the extension of data mining logics. In the remainder of this chapter, I reflect on how the concept of agency might be helpful in thinking about ordinary social media data mining.

Agency is a core concept in studies which seek to explore how cultures and societies are made, and how they might be made fairer and more equal. Agency is frequently opposed to structures in debates about which has primacy, which determines. Structuralist critics would argue that structures not only determine but also serve to restrict and oppress already disadvantaged groups in society. Marx's (1852) assertion that people are able to make history, or act with agency, but that they do so in conditions not of their own making, guides contemporary Marxist critics of capitalist structures which incorporate processes like data mining (for example Fuchs 2011; Hearn 2010, 2013). Some of the authors discussed earlier in this chapter arguably fall into this category. In contrast, others have stressed the capacity of individual human agents to make and shape their worlds. Others still have highlighted the dialectical relationship between structure and agency: structures shape and constrain human agency, but human agents act against, as well as within, them.

Jason Toynbee offers a useful summary of the ways in which critical realism understands interrelationships between structure and agency in his book about Bob Marley (2007), drawing on the work of Roy Bhaskar (1979). Within this framework, people are not seen only as components or

---

[4]The social analytics approach was first developed by Couldry and collaborators on the Storycircle project, to study how organisations use analytics to meet their goals. See http://storycircle.co.uk/.

effects of structure(s). Rather, they reproduce and occasionally transform society; they do not simply create it, as social structure is always already made. Bhaskar develops 'the transformational model of social activity', or TMSA, as a way of making sense of the relationship between people and society. Toynbee sums this up as follows:

> Society consists in relations between people, and as such is dependent on their activities which reproduce or (less often) transform society. From the other side, human practice depends on society; there can be no meaningful action without social structure. Crucially, this dependency on structure imposes limits on what people can do while never fully determining actions. In other words we have some autonomy as agents. (Toynbee 2007, p. 25)

Social theorist Derek Layder (2006) argues that rather than seeing dualisms (structure/agency, society/individual) as separate, opposing and locked in a struggle with each other for dominance, entities in dualisms should be thought of as 'different aspects of social life which are inextricably interrelated' (2006, p. 3), and so interdependent and mutually influential. Following Giddens, he understands agency as 'the ability of human beings to make a difference in the world' (2006, p. 4). The word 'agency', he writes, 'points to the idea that people are "agents" in the social world— they are able to do things which affect the social relationships in which they are embedded. People are not simply passive victims of social pressure and circumstances.' The action–structure dualism draws attention to the 'mutual influence of social activity and the social contexts in which it takes place' (2006, p. 4) and the ways in which social structures, institutions and cultural resources mould and form social activity. Similarly, Jeremy Gilbert (2012) insists on the need to develop a framework that accounts for both the creative agency of human actors and the ways in which structures shape and compromise social life. Gilbert proposes 'a perspective which can acknowledge the potency of both of these modes of analysis and the fact that they can both be true simultaneously'. In fact, he goes on to argue, 'I want to insist that we can't understand how capitalist culture works without understanding that they *are* both true' (2012, np). It is within this ever-present, dialectical tension between structure and agency, between control and resistance, that the questions at the heart of this book are situated. To ask what should concern us about social media data mining is to consider the extent of the dominance of structures, the possibility of agency, and the spaces in between. To combine perspectives which are critical of data structures with the perspectives of actors within data mining is to enrich understanding of data mining and datafication, as it means bringing together structural analyses with a recognition of individual agency in the context of these structures. This is what I attempt to do in this book. So far in this chapter I have focused on criticisms of structures. In the remainder, I outline three types of agency which are relevant to social media data mining: worker agency, user agency, and technology / agency relationships.

Worker Agency

In debates about social media and other forms of data mining, workers surface in contradictory ways. Sometimes, they are invisible, not seen at all. At other times, they appear to be all the same, an undifferentiated mass of powerful decision-makers. Elsewhere, occasionally, they are seen as victims, of the outsourcing of formerly professional labour to the amateur crowd or of 'function creep' (Gregg 2011), as the work of mining and monitoring social media data is added to their already full workloads. There are important exceptions to these patterns, which I discuss later, but first, I say more about these approaches.

Scholars writing about algorithmic power and culture sometimes write as if algorithms come into being on their own, rather than being brought into being by human actors tasked with the job of producing them. This can be seen, for example, in the work of Scott Lash (2007), already cited. With the concept of algorithmic power, Lash argues that data do not simply capture cultural life but serve to constitute it. We need to expand our understanding of the concept of power, Lash argues, to incorporate its increasingly algorithmic forms, because:

> They are compressed and hidden and we do not encounter them in the way that we encounter constitutive and regulative rules. Yet this third type of generative rule is more and more pervasive in our social and cultural life. [...] (It offers) pathways through which capitalist *power* works. (Lash 2007, p. 71)

A similar anthropomorphising of the algorithm can be seen in the works of Ted Striphas (2015) and his discussion of algorithmic culture. In an essay with that name, Striphas identifies that he follows Galloway in his use of the term, which he, Striphas, defines as 'the many ways we human beings have been delegating the work of culture—the sorting, classifying, and hierarchizing of people, places, objects, and ideas—to computational processes'. This move, he goes on to argue, 'alters how the category *culture* has long been practised experienced, and understood' (2015, p. 395). Algorithms are becoming increasingly decisive, argues Striphas, and are increasingly equipped with the task of imposing order on the mass of culture-data around us. In this context, social media platforms like Twitter and others 'bandy about in what one might call the algorithmic real' (2015, p. 407).

In these writings and others, the software engineers, data scientists and other workers who create algorithms or provide input into their development are not visible—algorithms appear to simply come into being, to spring from nowhere, and to take control. Arguably, culture is shaped not by algorithms but by the data workers who produce them, not to mention the people who implement algorithmic systems in specific organisational contexts. The production, implementation and *work* of algorithms involve humans and organisations and, as such, the work of determining and shaping cultural and social life is as much a fact of human as algorithmic agency. But to say that these actors have agency is not to say that they have complete power—structures constrain. Barocas and Selbst (2014) acknowledge this, distinguishing between 'decision-makers' and data miners, and so highlighting that it is often not data miners themselves who make decisions about how, where and what to mine, and so drawing attention to the differential distribution of power among those involved in data work.

Barocas and Selbst's discussion notwithstanding, in other work in which data labourers are acknowledged, they are sometimes undifferentiated. For example, in Andrejevic's writings, generalised terms are used—he writes about 'data miners' or 'sentiment analysts', and makes assertions about what this apparently homogeneous group of people do (Andrejevic 2011, 2013). Likewise, Hearn does not differentiate the 'feeling-intermediaries' she writes about, who 'structure feelings into profits for themselves and their clients' (2010, pp. 435–6, 428). Elsewhere, workers are victims of the spread of data mining and the associated expectation that workers in all sorts of institutional contexts will develop the expertise to mine data. Doyle (2015) writes about this in relation to the increasingly compulsory social media data mining that journalists are required to undertake in newsrooms. Another way in which workers are victims in relation to data work is through the crowdsourcing of data mining tasks through schemes like Amazon Mechanical Turk or Odesk (Caraway 2010), diminishing the pool of work available to professional workers. As will be seen in later chapters, some of these latter constructions of workers were relevant in my research, although I try to move beyond thinking of the workers I encountered as the passive victims of data mining, by considering their potential to act, or to have agency.

There are some exceptions to the patterns identified here. In some literature, the role data workers play in the production of datafied cultures is acknowledged, as are the differences among them. MacKenzie (2013) attends to the work of data mining, arguing that data scientists often embody the tensions and promises of data mining in ways that lead to troubling consequences. Gehl (2014, 2015) also writes about the work of data scientists, and Irani has studied the Amazon Mechanical Turk workers who do the micro-work of cultural sorting (Irani 2015a, b; Irani and Silberman 2013). Likewise, Gillespie (2012) writes about the content moderators whose job it is to review reported content on Facebook and decide whether it should be deleted, or what he calls 'the dirty job of keeping Facebook clean'. Gillespie argues that given the increasingly 'private determination of the appropriate boundaries of public speech' through these processes, it is imperative that we understand how such data work gets done, by whom and to what ends.

The writers discussed in the last paragraph draw attention to the labour involved in data mining, and the role workers play in data mining regimes. Focusing on workers is one way to consider the possibility of agency in relation to data mining structures, and so to think about whether the criticisms discussed in the first half of this chapter are applicable to forms of data mining taking place in small-scale organisations. This means studying not only data miners, but other data workers, such as account managers, marketers and strategists working for social media insights companies, and digital communications and public engagement managers within organisations which use the insights services of these companies. In subsequent chapters in this book, I draw on empirical research with a cross-section of these workers and reflect on relationships between data mining structures and worker agency. Now I turn to a discussion of another group of actors whose agency in data mining regimes we might consider: users.

User Agency

Users figure in debates about social media data mining in similar ways to workers, without an awful lot of recognition of their potential for agency. They are sometimes conceived as a group whose labour is exploited in the interests of the profitable accumulation of the social media platforms, so, interestingly, while workers' labour is frequently not acknowledged, the so-called labour of platform users receives more attention. Social media data mining exploits the unpaid labour of social media users, argues Christian Fuchs, for example, who draws on Marxist logic to argue that the activities of Web 2.0 users can be understood as an exploited form of 'cognitive, communicative, and co-operative labour—informational labour' (2011, p. 300, see also 2014). Thus producers of user-generated content are exploited labourers. According to Fuchs, the ecosystem of social media platforms, to use van Dijck's term, constitutes 'a commercial, profit-oriented machine that exploits users by commodifying their personal data and usage behaviour (web 2.0 prosumer commodity) and subjects these data to economic surveillance so that capital is accumulated with the help of targeted personalized advertising' (Fuchs 2011, p. 304). One manifestation of this view is the Wages For Facebook manifesto (http://wagesforfacebook.com/), launched in the spring of 2014 by US curator Laurel Ptak, which aimed to raise awareness of the invisible exchange that users make with social media platforms when they accept platform terms and conditions and then like, share, chat, tag and so on.

Another way in which users are seen as the objects of social media data mining is through their near-compulsory self-branding. In the digital reputation economy, it is argued, we see ourselves as brands, as saleable, exchangeable commodities. This leads us to self-brand, an activity described by Hearn as 'a highly self-conscious process of self-exploitation, performed in the interests of material gain or cultural status' (2008, p. 204). For Hearn, self-branding is a rational consequence of and response to the overarching conditions of advanced capital in which promotionalism is a dominant condition. In this age of digital reputation evaluation systems like Kred, Klout and PeerIndex, self-branding becomes a compulsory form of self-quantification.

As with discussions of data workers, users do not appear to have much scope for agency in these debates, constrained as they are by pervasive structural forces of data power. Compelled by social media logic, their involvement in social media leads to the exploitation of their own compulsive self-branding, in an effort to be visible. But there are other ways of thinking about users, not so much in literature about data mining but rather in other fields, which is helpful in considering the possibility of user agency in relation to data mining. These include the rich tradition of audience research in media studies and, within social media studies, some of the research exploring how social media users think and feel about their use of social media. Such studies enhance knowledge about what we can take social media data to be—feeling, self, identity, performance?—which in turn contributes to understandings of social media data mining. To give one example, in their research on Twitter users, Marwick and boyd (2010) identified extensive self-censorship, driven by their respondents' desire for balance in their tweets. Their respondents' tweets may therefore reflect more measured emotions than they actually feel, and what appears as opinion or sentiment may in fact be its performance. In another example, boyd (2014) highlights how young social media users play with their profiles to evade monitoring: they input false data about their age, location and relationship status, out of a belief that it is unnecessary for the platforms to request this information of them. This is not for purposes of deception—most of these users' social network friends are their friends IRL (in real life) and have a more accurate picture of them than Facebook or third party advertisers do. While modest in scale and not fundamentally disruptive, these studies show that engaging with users can tell us something about their agency. Taking this kind of approach is not to suggest a celebration of user power, but rather to propose that in the context of the critical literature on data mining discussed here, we should not forget that structures of data power are 'the continually produced *outcome* of human agency' as well as its 'ever-present *condition* (material cause)' (Bkaskar 1979, pp. 43–4).

Techno-agency

As well as thinking about the potential agency of particular actors (like workers and users) in relation to social media data mining, in order to undertake the project of this book it is also useful to draw on broader debates about relationships between technology and agency, and the types of agency that technological ensembles enable and constrain. As stated in the book's introduction, many sociological studies of technology have focused precisely on this intersection, on whether technologies can be appropriated, through acts of human agency, as tools of democratisation, despite their origins as apparatuses of power. Andrew Feenberg (1999, 2002) addresses this problem by asking whether human actors necessarily have to 'submit to the harsh logic of machinery', or whether other possibilities are open to us, in which human beings can act in ways that move us towards a better relationship between data mining and social life than the one described by critical commentators.

To account for such possibilities, Feenberg develops what he calls an anti-essentialist philosophy of technology. A core difference between Feenberg's project and my own, however, is that the criticisms of technology to which he objects are, in his terms, essentialist, in that technological developments are seen as essentially harmful; that is, harmful in essence. In contrast, and in part as a result of the passage of time, the criticisms I outline above are not based on assumptions about technology's essence, but rather focus on technology's co-option by forces of power. That distinction aside, Feenberg's argument about how to think about technology is relevant to this book, especially in relation to the 'ordinary actors' who are its focus. In *Questioning Technology* (1999), Feenberg suggests that at the heart of his proposed philosophy is a recognition of a fundamental difference among technical actors, which is the distinction 'between the dominant and subordinate subject positions with respect to technological systems' (1999, p. x). He describes this distinction as follows:

> ordinary people do not resemble the efficiency oriented system planners who pepper the pages of technical critique. Rather, they encounter technology as a dimension of their lifeworld. For the most part they merely carry out the plans of others or inhabit technologically constructed spaces and environments. As subordinate actors, they strive to appropriate the technologies with which they are involved and adapt them to the meanings that illuminate their lives. Their relation to technology is thus far more complex than that of dominant actors (which they too may be on occasion). (1999, p. x)

Feenberg then weaves together his understanding of these differential subject positions with his response to critical thinking about technology. For Feenberg, critical approaches offer 'no criteria for improving life within (the technological) sphere' (1999, p. xiv). He argues that critics ultimately agree with technocrats, because they all appear to believe that 'the actual struggles in which people attempt to influence technology can accomplish nothing of fundamental importance' (1999, p. xiv). Instead, he argues that change can come 'when we recognize the nature of our subordinate position in the technical systems that enroll us, and begin to intervene in the design process in the defense of the conditions of a meaningful life and a livable environment' (1999, p. xiv). User experiences are 'eventually embodied in technological designs', he argues, giving as examples women's claims with regard to the technologies of childbirth, the demands of AIDS patients that they be given access to experimental medication, and the use of a technology which was originally intended for the distribution of data, the internet, for democratic human communication. Citing Pinch and Bijker (1987), he points out that technological artifacts succeed where they find support in the social environment. This would seem to suggest that there is indeed scope for humans to act with agency in relation to technology.

I share Feenberg's concern that to dwell solely in the terrain of criticism does not offer 'criteria for improving life within that (technological) sphere' (1999, p. xiv). The danger with critique is that we stop there and do not move beyond it. Moving beyond critique, recognising and exploring what Feenberg call technology's 'ambivalence'—that is, 'the availability of technology for alternative developments with different social consequences' (1999, p. 7)—is an essential part of the project of thinking critically about technologies of power, including data power. So in this book, I explore what possibilities for agency exist in relation to data mining technologies, by asking whether there are forms of data mining, of social media and other data, that can enable ordinary people or ordinary organisations, and whether social media data mining can be used in ways that we find acceptable. In some of the chapters that follow, I discuss efforts to harness data mining's ambivalence, and to act agentically within the structures of data power.

Postscript on Agency: Acting Ethically in Times of Data Mining

Given that much of the critical work discussed in this chapter and later in the book focuses on the structures of data power, it seems sensible to frame the empirical research on which this book draws as an examination of the possibility of agency in the face of these structures. But what kinds of agency are implied within this framing? For some writers, agency is necessarily a reflexive practice. Couldry, for example, defines agency as 'the longer processes of action based on reflection, giving an account of what one has done, even more basically, making sense of the world *so as* to act within it' (2014, p. 891). Layder thinks like this too. He writes:

Social analysis must take into account the meaning that the social world has for the individual based on how the person understands and responds to their lived experience. The way people construe their social existence helps them formulate their plans and intentions. They make choices about the direction in which their lives should go on the basis of their experience. As such, persons are 'intentional', self-reflective and capable of making some difference in the world. (2006, p. 95)

For other writers (such as Bourdieu (1980)), agency is much less reflexive. It is exercised habitually, without thinking. In this framing, acting with agency is not necessarily reflexive or moral; it is not necessarily good. But for my purposes, I adopt an understanding of agency which incorporates elements of reflection, as this helps me explore technology's ambivalence, or the possibility of acting back against the problematic structures of data power. So, to some extent, and informed by Layder's assertion that action and agency are interchangeable, the version of agency that I mobilise here might be defined as 'acting ethically'. Thinking about the possibility of ethical action is also a way of responding to criticisms of data power. In making that claim, I am influenced, as in earlier work (Kennedy 2011), by J.-K. Gibson-Graham's *A Postcapitalist Politics* (2006), in which the authors speak back to criticisms of capitalism and neoliberalism by outlining 'myriad projects of alternative economic activism' (2006, p. xxi), which they define as both postcapitalist and ethical. Within such economic projects, the 'politics of possibility' which is at the heart of Gibson-Graham's vision emerges, as individual and formerly disempowered actors find new ways to exercise power, thus finding the grounds for a 'new political imaginary' (2006, p. xxi).

According to Gibson-Graham, alternative economic models exist at the most microscopic levels:

When a meal is cooked for a household of kids, when a cooperative sets its wage levels, when a food seller adjusts her price for one customer and not another, when a farmer allows gleaners access to his fields, when a green firm agrees to use higher-priced recycled paper, when a self-employed computer programmer takes public holidays off, when a not-for-profit enterprise commits to 'buying local', some recognition of economic co-implication, interdependency, and social connection is actively occurring. These practices involve ethical considerations and political decisions that constitute social and economic being. (Gibson-Graham 2006, p. 82–3)

Here, and in the more ambitious alternative economies models that they describe, they point to the *ethics* that pervade such practices. For Gibson-Graham, ethics is defined as 'the continual exercising, in the face of the need to decide, of a choice to be/act/think in a certain way' (2006, p. xxvii). Thus the projects which cause Gibson-Graham to be hopeful represent ethical economic practices, in which individuals and groups choose actions which embody the belief, captured in the motto of the World Social Forum, that 'Another world is possible'. Gibson-Graham's ideas about how to understand alternative economic models as part of a 'politics of possibility' might also be applied to data. Alternative ways of doing data mining could also be seen in this way, as human actors find new ways to exercise agency in relation to data, thus finding the grounds for a 'new political (data) imaginary' (2006, p. xxi). This book asks how we might solve the problems of data power that critique has highlighted, in part by seeking to identify data practices that might be seen as ethical acts of agency, and about which we might also feel hopeful.

CONCLUSION

The aim of this chapter has been to provide a framework for the empirical chapters that follow. I have sketched the main criticisms that have been levelled at data mining in the early phase of datafication. The most visible criticisms, I suggest, relate to concerns about the reduction in privacy and the expansion of surveillance that increasingly pervasive data mining brings with it. Spaces that feel private but are not are increasingly subjected to the surveillant gaze and, under such conditions, the character of surveillance changes. This can result in widespread practices of social sorting and discrimination, which in turn can be understood as forms of social control. Thus data mining acts in the interests of the powerful, facilitating the management of populations in ways that are increasingly opaque. Discrimination through data mining is unintentional as well as intentional and, through their discussion of the moments at which such discrimination can occur, Barocas and Selbst (2014) draw attention to another set of criticisms, relating to the methodological and epistemological issues that data mining raises. And, just as data mining can exclude populations from its algorithmic calculations because of its methodological particularities, so it can be exclusive in another way, in terms of who has access to data mining tools and technologies, and the skills needed to participate in data-driven operations.

All of these criticisms represent things that should concern us. But as datafication takes hold, and the mining of social media and other data becomes more ordinary and everyday, we need to ask whether these criticisms, developed in relation to the high-profile, spectacular data mining undertaken by powerful entities like governments, security agencies and the social media platforms themselves still hold as we lower our sights and focus on data mining 'activities in the daily round' (Silverstone 1994).

The terrain of social media data mining is no longer occupied only by profit-driven mega-corporations and suspect governmental and security agencies. To understand social media data mining as it becomes ordinary, we need to consider these actors and, as Ruppert et al. put it, attend to 'specific mobilisations which allow the digital to be rendered visible and hence effective in particular locations' (2013, p. 31).

Broadening our understanding of ordinary social media data mining practice—who does it and what its consequences are—we also need to consider whether there are forms of data mining that are compatible with the interests of subordinate actors within technocratic systems. Like Feenberg, I am sympathetic towards Marxist-influenced criticisms of the ways in which technologies are harnessed in the interests of the powerful, but also like him, I want to consider whether it is possible to act with agency in relation to social media data mining and so harness technology's ambivalence. As stated in the introduction, if we are committed to a better, fairer social world, we need to consider these possibilities. The perspectives introduced in the second half of this chapter might make this possible, as they focus on questions of agency and ethical action. Drawing on these perspectives, we can think about how data mining is shaped by actors within specific organisational contexts, how users manoeuvre within data mining structures and whether there is room for agency in the face of data power.

The ideas discussed in this chapter form a scaffold for interrogating ordinary and everyday forms of social media data mining, to adapt a term used by van Dijck and Poell (2013) to describe the economic structures that underlie and maintain social media logic. Scaffolds are not always visible, but what is built upon them would fall apart if they were not there. Jason Toynbee writes that critical realism's approach to the structure/agency dialectic, which he summarises at the beginning of his book on Bob Marley, influences the rest of it, sometimes implicitly, like an invisible scaffold. In other words, his book is built on the assumption that 'social structure has a real existence, that causes things to happen through enabling and constraining the actions of actors' (2007, p. 34), but he does not repeat that point throughout his book. I make similar claims about this book. It emerges from the debates considered here, both the criticisms of data power and the different ways of thinking about agency and ethical action. All of these ideas contribute to the book's scaffold. They do not appear repeatedly in the chapters which follow, because sometimes these chapters focus on other things that should concern us about social media data mining which are not discussed here, and others still that might not concern us. But they inform—and scaffold—the chapters that follow.

CHAPTER 4

Public Sector Experiments with Social Media Data Mining

INTRODUCTION

The bold assertions about what the analysis of social media data might tell us with which I introduced the last chapter exist alongside funding cuts in the UK's public sector (Lowndes and Squires 2012). Significant changes are under way in public sector organisations in the name of austerity, and social media data mining is seen to provide a potential solution to the problem of diminishing resources in city councils, museums and other cash-strapped public bodies. In the context of ubiquitous rhetoric about the potential of big data, widespread datafication and austerity measures, public sector organisations have started to think about how they might harness social media data mining as it becomes ordinary (and they are not alone in believing in the potential of data mining to create efficiencies and deliver valuable information about the people they want to engage).

In this chapter, I turn my attention to how social media data mining is being used by public bodies, and what, if anything, should concern us about public actors' uses of these methods. To answer these questions, I draw on action research I undertook in collaboration with Giles Moss, Christopher Birchall and Stylianos Moshonas in 2012 and 2013,[1] as social media data mining started to come onto the horizons of this

---

[1] Other publications discussing this research include Kennedy et al. (2015), Kennedy and Moss (2015), Moss et al. (2015).

© The Editor(s) (if applicable) and The Author(s) 2016   67
H. Kennedy, *Post, Mine, Repeat*,
DOI 10.1057/978-1-137-35398-6_4

sector. Working with city-based public organisations, we attempted to find out about existing uses of social media data mining, experiment with these methods to evaluate their potential use and reflect on their normative consequences. We were motivated to do this by a number of factors. First, we wanted to explore whether it was possible for such methods to be accessible to those with limited economic means and so to circumvent the threat of a new digital divide based on differential levels of data access. In this way, our research was born out of some of the criticisms of these methods made by boyd and Crawford (2012) and discussed in Chapter 3. At the same time, we were interested in examining ways in which resource-poor groups who want to use social media data mining for the public good might be able to do so. In the context of the austerity measures at the time, public sector organisations fall into this category; they, like many others, are in danger of ending up on the wrong side of the divide. So in this sense, we were interested in exploring the question of whether it is possible to submit the machinery of social media data mining to the public good. Finally, influenced by the idea that methods enact realities (Law and Urry 2004) also discussed in Chapter 3, we wanted to explore the ways in which 'the public' is enacted through social media data mining. Researching this subject was important, we believed, because the types of publics that emerge through such methods can become the basis for decision-making about things like the provision of services. In this sense, we wanted to reflect critically on the construction of publics through social media data mining. So we approached our action research both open to the potential of these methods and aware of the problems that they bring.

Realising these aims proved challenging. For all of the contextual reasons mentioned above—big data rhetoric, datafication, austerity measures and efficiency drives—our partners were committed to using social media data mining methods in order to find data, identify and take action and get results, and we found ourselves playing a role in producing this desire, our critical perspective notwithstanding. We found that it was difficult to enact our commitment to both the potential *and* the problems of data mining methods, as our feelings of responsibility towards our partners to produce results sometimes eclipsed our attention to critical inquiry. Hammersley (2002) suggests that sustaining an equal balance between action (understood as potential here) and research (understood as problems) might always be difficult in action research, but I argue that the difficulties we encountered were as much about social media data mining as action research. This is because using social media data mining is also motivated by a will to produce results. Bringing together Porter's discussion of trust in numbers in a book of that name in the 1990s (Porter 1995) with Grosser's more recent work on the ways in which the metrification of sociality on social media platforms creates a desire for more and more metrics, I describe this will for results as a 'desire for numbers', something that was prevalent across the sites of my research. I develop this notion of a desire for numbers here and in later chapters, starting here by charting its influence on our collaborations with our public sector partners. In our research with them, public sector workers' acknowledgement of the limited capacity of social media data mining, for example to fully 'represent' publics, was encouraging, but nevertheless, the predominance of this desire for numbers meant that we had limited opportunities for reflecting with our partners on the ways in which data mining methods make and shape their objects. In this chapter, I show that despite the research team's awareness that our interventions in the data mining process played a role in constructing particular publics, we were not able to create spaces in which to think critically with our partners about these issues.

These difficulties notwithstanding, public sector social media data mining is motivated by aims such as greater understanding of public opinion or greater inclusion in public processes of groups who might otherwise be excluded from them. In these ways, public sector social media data mining aims to serve the public good. Recognising that concepts like the public good, engagement and empowerment can be problematic, I nonetheless argue that it is empirically inaccurate (to borrow a phrase from Banks (2007) on how to theorise cultural work) to understand public sector engagements with data mining as *only* problematic. Taking seriously the possibility that some forms of data mining might serve some form of good is a necessary part of a commitment to exploring whether it is possible to exploit technology's 'ambivalence' or 'the availability of technology for alternative developments with different social consequences' (Feenberg 1999, p. 7). Before discussing these points in detail, the next section provides a brief discussion of the relationship between data mining methods, public bodies and public engagement, and our action research approach.

ACTION RESEARCH AND THE PRODUCTION OF PUBLICS

Knowing and Forming Publics

As my collaborators and I have observed elsewhere (Moss et al. 2015), public sector organisations' remit to understand and engage the complex and changing publics that they exist to serve becomes more difficult in the context of austerity. Local governments adopt various methods to connect with their local publics, such as consultations, citizens' panels and consumer-feedback mechanisms (Barnes et al. 2007), but methods like social media data mining appear to offer a cost-efficient new way for these organisations to know, understand and engage their publics. Analysing social media data may also provide them with access to formerly excluded groups.

As noted, social media data mining not only serves as a means to know and engage publics, but also to bring publics into being. Gillespie (2014) argues that the representations of the public generated through such methods do not simply mirror the public 'out there' but also construct it in particular ways. He argues that we should therefore ask, 'how do these technologies, now not just technologies of evaluation but of representation, help to constitute and codify the publics they claim to measure, publics that would not otherwise exist except that the algorithm called them into existence?' (Gillespie 2014, p. 189). These are important questions, but when it comes to representing publics, they are not new. A number of writers have noted that 'the public' comes into existence partly through technologies of representation, such as opinion polls (Moss et al. 2015; Peters 1995). Given this, and when we consider how vast and differentiated publics are, there may be some value in the representational work that social media data mining can do in making publics knowable (Anstead and O'Loughlin 2014). Indeed, it could be argued that it is not the production of publics through data mining that is problematic, since publics are regularly produced in these and other ways, but *how* they are constructed and with what social effects, because social media data mining methods represent just another way of bringing publics into being. So when Gillespie (2014) describes the publics that emerge through these methods as 'calculated publics', contrasting these with 'networked publics', which he describes as 'forged by users' (2014, p. 189), it could be argued that the problem with the 'calculated publics' produced by algorithms is not that they are productions per se, but that they are typically limited, passive and formed with narrow commercial purposes in mind. A further problem is that, unlike more conventional means of representing publics, we know little about the specific ways the public is constructed through social media data mining methods because of its black-boxing.

In her research into politicians' uses of social media, Roginsky found that in social media workshops for Members of the European Parliament (MEPs), there was a tension and confusion between what she calls 'company-selling-products' approaches and 'engaging-the-public' approaches, as companies focused on the former are often called upon to advise the latter (Roginsky 2014). This is the dilemma at the heart of efforts to do good with data mining: the same tools can be used for both democratic and commercial gain. To date, data mining has primarily been used as a way for corporations to know, profile and discriminate among members of the public, and tools and techniques have largely been developed in corporate contexts. Not only are corporations dominating the field but, given that the aims of data mining are distinct across sectors, available tools may not travel well and be appropriate for all (Baym 2013). The publics produced through data mining may well be limited, as noted above, divided as they are into market segments not aware of themselves as a collective subject. The question that this leaves us with, then, is whether the calculated publics brought into being through algorithmic practices can be different kinds of publics, fuller, more active and reflective. This might be seen as a use of social media data mining for the public good. To explore this question and to understand how these methods shape the publics that they are mobilised to understand, we undertook the action research project described below.

Action-Researching Public Uses of Social Media Data Mining

In 2013, we undertook action research for a period of 6 months, working with two city councils and one city-based museums group in the north of England to explore how data mining might help them to understand and engage their publics. In naming our approach action research, I point to our intention to work collaboratively with our partners, to engage in both research/inquiry and action/intervention. As a number of action researchers have observed, at the core of this approach is an intimate relationship between scholarly inquiry and practical, political activity and intervention, 'such that the focus of inquiry arises out of, and its results feed back in to the activity concerned' (Hammersley 2002; see also Freire 1970; Reason and Bradbury 2001). In this model, action is assumed to be good, ethical, emancipatory, as in the concept of agency that I mobilise in this book.

Central to much debate about action research is the question of whether it is possible to sustain both action and research in equal measure. While some subscribe to the Greek privileging of research, or theoria, over action, or praxis (for example Polsky 1971), Hammersley argues that the subordination of inquiry to action is more common. This is because, as Reason and Bradbury put it:

> the primary purpose of action research is not to produce academic theories based on action; nor is it to produce theories about action; nor is it to pro- duce theoretical or empirical knowledge that can be applied in action; it is to liberate the human body, mind and spirit in the search for a better, freer world. (Reason and Bradbury 2001, p. 2, cited in Hammersley 2002, np)

Both models—the subordination of action to research and the subordi- nation of research to action—are legitimate, in Hammersley's view. In our experience, I argue below, the subordination of research to action was a result of these common tensions in action research but also the desire for numbers that engagement in data mining produces, because the produc- tion of results (whether social change or digital data) is the ultimate aim of both. I say more about this below.

Prior to the action research, Giles Moss and I carried out interviews in five public sector organisations in northern cities in England (two city councils, two museums groups and a regional tourism organisation), to examine whether and how they were using social media data mining in relation to their public engagement objectives. This formed part of a larger study of the impact of digital developments on public engagement (reported in Coleman et al. 2012). Through these interviews, we identi- fied that organisations were using some tools to monitor their social media activity, such as TweetDeck (an application for managing, organising and tracking Twitter accounts) or Museum Analytics (a platform which produces summaries of social media activity for museums, mentioned in Chapter 2). These tools were not used systematically and most organisa- tions were keen to expand their use of data mining methods. In addition to these interviews, we developed our understanding of how the tourism organisation was mining social media through an undergraduate student's paid internship and participant observation there.

Given interviewees' expressed enthusiasm to experiment with more data mining methods and the limited availability of adequate resources to do so, we invited them to collaborate with us to explore a wider range of tools than they already used. Two city councils and one museums group accepted the invitation. Others declined. One of the museums groups did so because staff believed that their existing approach of simply keeping an eye on discussions on city forums and on their own Facebook and Twitter accounts were sufficient for the small scale of their operation. For them, approaches which relied on human rather than technical agency seemed more appropriate to their public engagement needs than data mining methods.

Then, through consultation with our partners, and with an expert from a commercial social media insights company who had extensive experience of working with the public sector, we identified some free or affordable tools with which to experiment, many of them discussed in Chapter 2. These included: NodeXL (http://nodexl.codeplex.com/), a freely available social network analysis tool; Gephi (https://gephi.org/), an open source tool for network visualisation; and IssueCrawler (www.issuecrawler. net), a free tool which identifies issue networks (that is, networks linked by interest in specific issues, rather than social networks). We also used DataSift (http://datasift.com), a commercial, online tool to harvest data from a variety of social media platforms at low cost (at the time of writing, 20 US cents per unit, which can include up to 2000 tweets). NodeXL accesses a limited number of platforms, one at a time, so adding DataSift to the toolset allowed us to generate multi-platform datasets which could be exported into the other applications. We aimed to explore whether these tools would enable our partners to identify significant yet hitherto unknown influencers within their target communities with whom to engage, which they identified as one possible use of social media mining.

We also trialled two of the commercial social media insights tools mentioned in Chapter 2, one at the request of one of our partners, and the other as a comparison: Meltwater Buzz (http://www.meltwater. com/products/meltwater-buzz-social-media-marketing-software/) and Brandwatch (http://www.brandwatch.com/). Together, all of these tools enabled us to carry out investigations that covered some of the major categories of data mining: social network analysis, issue network analysis, exploratory content analysis and visualisation of the resultant datasets. The toolset was expanded as the project progressed, because the free tools were experienced as difficult to use, because one of our partners wanted us to experiment with a commercial tool produced by a company from which they licensed a traditional media monitoring platform and because, during the course of the research, it became apparent that less data mining was taking place than our interviews implied. This latter point suggests that our interviewees may have wanted to create the impression that they were more involved in social media data mining than was actually the case, an indication of the power of big data rhetoric to persuade people that this is what they *should* be doing.

Our inclusion of commercial tools within the project toolset indicates a number of things about the action-imperative of action research and data mining. It points to some of the problems our partners encountered using NodeXL, Gephi and IssueCrawler because of what they saw as the tools' complexity; I say more about this later. More importantly, our move to include commercial tools demonstrates the power of the desire for numbers and the subsequent difficulty we encountered balancing research with action, or problems with potential. In agreeing to collaborate with us, our partners hoped to find some data; they used data mining methods because they were motivated by a desire to get results, a desire for numbers. To enable this, we used commercial tools, which we hoped would be more effective than free tools were proving to be—partners were struggling to use them, and we were struggling to find data with them. This move constructed us as intermediaries between the tools, their developers and our partners. It compromised our initial intention of experimenting with free technologies which enable people to access digital data despite limited economic means, as well as moving us closer to the action and further from the research of action research. It also meant that we, like the methods and tools, played a role in constructing the publics that resulted from our experiments and that these were produced with tools developed primarily for commercial purposes.

Over six months, we worked with our contacts in the communications teams of our partner organisations to experiment with these tools. Meetings were more frequent with the museums group than the other two partners, primarily because of the greater availability of staff therein. Thus what started as a collaborative and participatory project became less so for two partners, who subsequently had less opportunity to develop data mining skills than we had all hoped. To compensate, for each partner, we produced a report summarising what had been found about their organisations through our experiments. These reports were intended as indications of what is possible with social media data mining, rather than as comprehensive accounts. We also produced a 'guide to tools', which we shared with all partners, to enable them to continue experimenting after the end of the project. In it, we pointed out that there is much that is not known about how tools work, that choices about how tools are made shape the data they produce, and that access conditions constantly change. This represented an attempt to address the ways in which data mining methods shape the things they seek to understand and to highlight some of the problems with these methods. Sharing this guide was intended as a social intervention, to enable public sector organisations to engage in data mining. But producing instructions on how to use tools constructed us not simply as intermediaries between tools and partners, but as *advocates* for the methods and tools discussed. As Law et al. (2011) point out, this is how methods work—they need advocates in order to be adopted. Our advocacy of data mining methods overshadowed our attempt to create spaces in which to discuss problems—there was one page of 'problems' compared to 46 pages of instructions in our guide to tools. Here again, our interventionist intent eclipsed our research intent as we responded to our partners' desire for numbers.

To further compensate for two partners' limited engagement in the experimental phase of the research, we ran a training workshop at the end of our project, which offered partners a hands-on opportunity to experiment with tools. It was attended by 13 representatives from a range of departments in our partner organisations. In both the workshop and the guide, we introduced more freely available social media data mining tools than those we explored in the action research in order to equip our partners with quick-and-easy data mining methods, as our research had shown that it was difficult for them to dedicate time and effort to using more complex tools. These included some of the 'free-and-easy' tools discussed in Chapter 2, such as Social Mention (http://www.socialmention.com/), which aggregate content from social media sites and produce simple statistics, such as numbers of comments on a given topic.

There are other methods and tools that we could have used to mine and analyse social data, all with distinct strengths, limitations and affordances. We were led in our choices by our own knowledge, the advice of experts and the requirements of our partners. Tools needed to be free or cheap given our partners' resource constraints, and the kind of analysis they facilitated was also an important consideration, as most partners had informed us that they were already doing some analysis, and we therefore sought tools which would allow them to expand on their existing practices. These criteria changed as the research progressed. In the following sections, I discuss our experiences of working with our chosen tools, highlighting potential and actual public uses of social media data mining, and the research team's role in facilitating such uses.

SOCIAL MEDIA DATA MINING FOR THE PUBLIC GOOD?

Uses of Social Media Data Mining

Through the interviews we undertook in five public sector organisations in 2012, we found that social media and data mining tools listed in (Table 4.1) below were being used.

These tools were used with varying degrees of regularity. All organisations had accounts across several platforms and none maintained them all regularly. In some organisations, one platform was used more regularly and others more sporadically; in other cases, a particular Twitter feed, for example, would be especially active, whereas other feeds were less frequently utilised.

Interviewees felt that their organisations were under-resourced, which meant they dedicated limited resources to social media engagement and analysis. Some were succeeding in managing and monitoring social media despite limited resources, because of their small-scale or autonomous structure, whereas others wanted to do more with the limited resources that they had. Some organisations passed on quantitative data to funders or senior managers, using tools such as Museums Analytics to produce summaries of social media activity such as new page likes, posts and comments on Facebook, and new followers, tweets or mentions on Twitter. Interviewees did not know what, if anything, became of this data, and some of our interviewees questioned the extent to which quantitative data could measure 'real' engagement, believing instead that more emphasis should be placed on qualitative data. Other interviewees felt that it was difficult to persuade senior managers of the value and benefits of social media monitoring. To demonstrate the limited understanding of social media culture that existed among senior colleagues, one interviewee jokingly told us that his manager thought LinkedIn was called LinkedUp.

Table 4.1 Social media platforms and data mining tools used by public sector organisations in which we carried out interviews

Social media	Data mining tools
• Twitter	• Google Analytics
• Facebook	• Museum Analytics
• YouTube	• Google Alerts
• Flickr	• Rate This Page functions
• AudioBoo	• Postling
• Instagram	• Hootsuite
• Get Glue	• One Riot
• Pinterest	• Site Improve
• Foursquare	• Storify
• blogs	• Crowdbooster
	• iTunes download monitoring
	• internal Facebook/Twitter tools

One of the councils monitored social media for both customer services and reputation management purposes, and identifiable actions sometimes resulted from this kind of monitoring—for example, changes to the structure of their website as a result of complaints about difficulties in finding certain information.

On the whole, social media data mining methods were not used strategically by any of the organisations. These interviews were carried out, after all, at a time when the public sector was only just becoming aware of data mining. Monitoring of social media occurred across the organisations, but it tended to be informal and unsystematic. Some saw a gap here, believing that social media data mining would allow the organisations to pick up on and respond to local issues and concerns before they escalated, while others seemed less convinced of this need. Raising concerns about the rules of engagement, some interviewees felt that there was no need to go 'fishing for debates or conversations' and expressed anxiety about 'delving into someone else's space'. These views showed that some of our small group of interviewees exercised caution with regard to what it is and is not acceptable to mine and monitor on social media.

Nonetheless, all organisations identified further social media data mining activities in which they could engage. Through our initial interviews and a further 13 that took place during the action research, we identified these ways that participants imagined that data mining could be used by their organisations in the future:

- Measure engagement
- Identify key influencers
- Analyse feedback
- Monitor publicity
- Identify what matters to publics
- Co-design policy.

In relation to the first point, measuring the public's connection and engagement with the organisation, some participants felt social media data mining could be used to measure the circulation and reach of organisational messages across local networks and to explore the level of public interest and engagement generated. Another potential use of social media data mining was to identify key influencers, intermediaries and networks with which to engage. Data mining methods could allow the organisations to identify local networks of which they are unaware, 'finding out', as one interviewee put it, 'about people talking about issues that we don't know about' (Council Elections, Equalities and Involvement Officer). Having a better understanding of the make-up of local networks could help the organisations to spread messages more widely and effectively, including to 'hard-to-reach' groups that may not be accessible via conventional channels.

Social media data mining could also be used to manage and analyse enquiries and feedback from the public about services and initiatives, alerting organisations to particular issues that arise, while the aggregation of data could be used to detect trends, positive and negative. Likewise, they could be used to monitor relevant publicity, or what the public, key influencers and intermediaries may be saying about the organisations, not through direct feedback but in other channels. Interviewees described how social media data mining could be used as part of a proactive reputational management strategy, where the organisation mines social media in order to capture and publicise positive sentiment and comments about the organisation. It may also be used as part of a more defensive strategy, where organisations seek to manage reputational risks.

Others proposed that data mining methods could be used to identify and analyse what the public is saying about matters of local concern, capturing public conversations and views that may not otherwise find their way to the organisation via conventional channels, to inform organisational decision-making. One interviewee, a Council Elections, Equalities and Involvement Officer said:

Is there a way we can capture that information and add it into the mix that this is what people in the city think? Yes, you may not necessarily capture all of it, and you can't capture conversations people are having face to face that you don't know about; but you can try and capture a bit more.

More ambitiously, some participants flagged the possibility that social media data mining methods could be used to involve the public in the 'co-design' of policy. This would entail moving from a top-down model of engagement to a more participatory approach, where the organisation involves the public in policy formation. One interviewee said:

> I think what we're trying to move more towards, rather than us saying we're going to introduce this new initiative, or we're thinking about closing a building, or something like that, we need to get people involved at an early stage in that co-design, get public opinion about what is it they want from public services, how do they want them shaping, how do they want to design them with us. [...] Tools like this would be really helpful in supporting us do that. (Council Intelligence Manager)

Our six-month action research project, intended as an experiment in what was possible with social media data mining in small-scale, cash-strapped public sector organisations, was not going to be able to trial all potential uses, so we tested, on a small scale, the potential of digital methods to investigate public engagement with the organisations, as well as to examine broader local networks and conversations. In the next section, I discuss some experiments which our partners felt were beneficial, before moving on to some of the problems that we encountered with mining publics in social media. Across all of these examples, I highlight the role played by a pervasive 'desire for numbers'.

Understanding Publics, Desiring Numbers

To explore publics' engagements with our partner organisations, we analysed the volume of contributions linked to them through social network analysis of particular data sources, such as Facebook pages and groups or Twitter accounts and hashtags. Our participants thought that the results generated by these experiments were useful, with the network analysis of institutional Facebook pages or Twitter accounts through NodeXL proving especially beneficial. Network analysis provided valuable insights into active accounts, pages and groups, and the level of interaction generated between contributors. Figure 4.1, for example, shows a representation of

Figure 4.1   Networks of participants in relation to 1 month of activity data from the Facebook of one of the partners

1 month of activity data from a Facebook page run by one of the partners. The nodes in the diagram represent contributors to the Facebook page and the lines represent interactions between them, such as replies to posts. The contributors are differentiated according to their influence in the network (indicated by node size) and the communities of closely linked nodes to which they belong (indicated by colour). The names of the contributors have been removed from the diagram to retain anonymity for the purposes of publication, making it hard to see what we actually found, but giving an indication, nonetheless, of how these findings were represented visually. Using these methods, participants could identify key contributors, who is linked to whom, and particular platforms could be compared in terms of the degree of public engagement and activity they generate.

We used Gephi, NodeXL and IssueCrawler, as well as easier-to-use commercial tools to analyse local networks, and this also generated relevant data and insights. Using these methods, we located important groups of which our partners were previously unaware. For example, one of the maps generated by IssueCrawler (which shows how content is shared and connected online) identified a transport organisation which was communicating actively about the museum group's cultural events and so could be a beneficial contact for them. Our analysis also identified local online forums which were active in terms of user contributions about our partners, but about which, again, our partners were previously unaware. The commercial tools produced an analysis of contribution channels, indicating the proportion of the contributions from each channel that was harvested by each search, thus identifying key channels for each partner (see Figure 4.2).

The data generated by the tools were met with enthusiasm by some participants, especially when presented in visually appealing charts and graphs. A sense of amazement was expressed by participants who read the reports we produced and by those who attended the workshop. One research participant said that:

> I think I had a lot of confidence in the numbers. I think I was amazed by how deep a lot of these tools could go. [...] I think they're very clever. It was amazing how much you could drill into this. (Council Web Team Officer, Customer Services Department)

Across many sites of my research, as later chapters demonstrate, such enthusiasm and amazement was common. I characterise these responses as indicative of a pervasive desire for numbers, which has several origins, not all of which are specific to the phenomenon that I discuss in this book, the becoming-ordinary of social media data mining. In the 1990s, Theodore Porter identified what he described as a widespread 'trust in numbers' in a book of that name (Porter 1995). In *Trust in Numbers*, Porter asked the question: 'how are we to account for the prestige and power of quantitative methods in the modern world?' He argued that numbers appeal in various ways. They are familiar and standardised forms of communication and, as such, they can be understood from 'far away'—that is, by people distanced from and unfamiliar with the topic to which the numbers refer. Quantification is a technology of distance, he argues, and Kate Crawford (2013) makes a similar point about big data: they allow us to view phenomena from afar, but this means that we miss the detail that can be observed on closer scrutiny.

Numbers are universal, argues Porter—they can be shared across cultures, nations and languages. They are impersonal and therefore objective, and this in turn minimises the need for 'personal trust'. With these

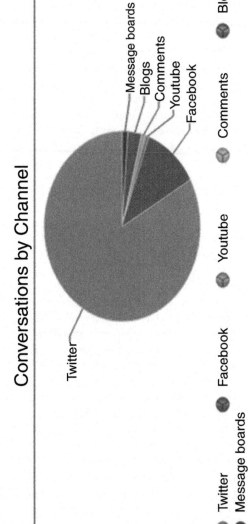

Figure 4.2 Key contribution channels for one particular search, identified by one of the commercial tools

words, Porter identifies *why* a trust in numbers needed to be cultivated at a particular moment in history marked by the growth of a centralised and bureaucratic state, managed by officials 'who lack the mandate of a popular election'. These officials needed processes which appeared objective, and this explains the turn to numbers and quantification, which had not always been recipients of the trust that Porter identifies. As Porter notes, 'a decision made by numbers (or by explicit rules of some other sort) has at least the appearance of being fair and impersonal' (Porter 1996, p. 8). The growth of quantification can thus be seen as an attempt to develop measures of impersonality at a time when trust was in short supply; it responded to uncertainty in modern life and modern organisations: 'objectivity lends authority to officials who have very little of their own', notes Porter (Porter 1996, p. 8). Porter argues that this faith in numbers and the credibility that is assigned to them is 'a social and moral problem'. He cites Horkheimer and Adorno's criticism in *Dialectic of Enlightenment* (1948) to substantiate this assertion: positivist science and its dependence on numbers replaced 'the concept with the formula, and causation by rule and probability', they noted (in Porter 1996, p. 18). Quantification is a process of managing the world, ordering it, not understanding it: what is lost when numbers dominate are the understandings that qualitative sensibilities help us to generate, as Baym (2013) writes in relation to big data.

In times of datafication, quantification is more widespread than ever, as a direct result of the ubiquity of social media. José van Dijck provides a full and convincing description of the role of social media platforms in the datafication of aspects of life that were previously understood qualitatively:

> With the advent of Web 2.0 and its proliferating social network sites, many aspects of social life were coded that had never been quantified before—friendships, interests, casual conversations, information searches, expressions of tastes, emotional responses, and so on. [...] Facebook turned social activities such as 'friending' and 'liking' into algorithmic relations (Bucher 2012; Helmond and Gerlitz 2013); Twitter popularized people's online personas and promoted ideas by creating 'followers' and 'retweet' functions (Kwak et al. 2010); LinkedIn translated professional networks of employees and job seekers into who communicated with whom, from which location, and for how long digital interfaces (van Dijck 2013b); and YouTube datafied the casual exchange of audiovisual content (Ding et al. 2011). Quantified social interactions were subsequently made accessible to third parties, be it fellow users, companies, government agencies, or other platforms. (van Dijck 2014: 198–9)

What's more, the rhetoric that accompanies datafication of this kind encourages more and more trust in numbers—Anderson's claim, cited earlier, that 'with enough data, the numbers speak for themselves' (2008) exemplifies this rhetoric. Such claims exist alongside the conditions of austerity and related demands on public sector organisations referenced above, and what Granovetter calls 'mimetic isomorphism', or the tendency of organisations to imitate each other ('whatever everyone is doing you do, because otherwise you look like you're not a modern firm' (1985, quoted in Perlin 2011, p. 40)). It is not surprising, then, to encounter a desire for numbers in such contexts.

But a trust in numbers and a desire for numbers are not the same. Back in the public sector organisations, some participants were more measured in their responses than those cited above. Despite Anderson's claim, what conclusions can be drawn from mined data was not always clear. One participant explained that while the data generated by the project provided useful contextual information, it was hard to identify its specific value and draw practical conclusions from it. One participant said:

> I can see it's very useful as providing additional context but in terms of actually being something that I could say, 'Ah right well, with this project, this is the way I need to go to get to where I want to be,' it doesn't do that and I think that would be a general view in the directorate. (Head of Strategic Planning, Policy and Performance, Council City Development Department)

Another workshop participant explained how she wanted to go beyond basic quantitative information. She said:

> You can't look at the data on face value and say 'Well, he generates lots of mentions and lots of content so he's a key influencer' because if you actually go into that site and have a look at some of the content that's generated, it's not particularly qualitative and actually he doesn't have that great a following on that particular website.

She went on to say, 'So you need to understand who the people are, what they're saying and how that's received by the audience as well and that bit is missing' (Councils Communications Officer). With these words, this participant acknowledged the limitations of numbers, and the need for qualitative as well as quantitative data.

One important difference between our contemporary datafied times and the historical examples on which Porter draws is that today, *distrust* in numbers exists alongside trust (as evidenced in Huff's *How to Lie with Statistics* (1954) and elsewhere). In research I have carried out about how people interact with data visualisations, which I do not discuss in this book, I found widespread distrust in data among participants, especially if reported in media that were also not trusted (Kennedy 2015). In a review of Porter's book on the online forum *Issues in Science and Technology* (http://issues.org/), Ravetz (1996) pointed out that the 'dirty truths' of science live alongside the field's 'public face of perfect, impersonal objectivity as guaranteed by its numbers'. Focusing specifically on environmental risk assessment, Ravetz argued that:

> there is now a vigorous public debate about the numbers, and no one is in any doubt that values and assumptions influence the risk calculations. Yet the field flourishes and is actually strengthened by the admission that it is not covering up hidden weaknesses. (1996, np)

To make sense of this contradiction, I turn to Benjamin Grosser's argument that the entanglements of metrics and sociality on social media platforms like Facebook transform the human need for validation of our personal worth into what he describes as an 'insatiable desire for more' (more likes, more followers, more retweets, more shares). In this context, quantification is a way of evaluating whether that desire has been fulfilled (Grosser 2014). Grosser proposes that the quantification of sociality which results from social media metrics creates an incentive to increase likes, comments, friends and so on. He describes this as a 'transformation, within the confines of capitalism, into this desire for *more*' (2014, np). Capitalism's own desire for growth leads to a culture which constantly audits whether the desired growth has been achieved, and this is experienced at the level of the individual or the organisation as a need to excel within the audit's parameters, he writes. This constant auditing deploys quantification to enable its very existence and value subsequently becomes attached to quantities, or numbers.

What I witnessed across a number of research sites, especially those discussed in this and the following two chapters, was a combination of trust, distrust and desire for numbers. In the public sector organisations, there was evidence of participants' hesitation in relation to what social media data mining can do and offer, as seen in the examples of the Head of Strategic Planning and Council Communications Officers above. In other examples, some participants acknowledged that there are significant exclusions or absences in social media data—the extent to which all social groups are represented in data produced through social media analytics is clearly a central concern for public organisations which aim to serve and represent the public as a whole, rather than one particular group. Those who do not participate on social media platforms will not be reflected in the data, while those who are not active contributors will be less visible than the 'super-participants' (Graham and Wright 2014) who dominate social media spaces, noted some participants. One participant said, 'I think trying to use social media doesn't cover everybody … there are certain audiences that you're more likely to reach. […] You're not going to get the total picture from any of (the tools)' (Council Elections, Equalities and Involvement Officer). But, as I argue later in this chapter and in the book, despite such acknowledgements, the dominance of a desire for numbers limits our capacities to convert such thoughtful reflections into a fuller critique of the limitations of, or problems with, social media data mining. I say more about this in the next section.

Should these instances of public sector social media data mining concern us? The examples discussed here, actual and potential, are modest and small in scale: a museums group identifying organisations to whom they can promote their cultural events, local councils identifying which social media channels are effective for them, or developing their understanding of how members of their publics are linked to each other in online networks. As we lower our sights to focus on ordinary social media data mining (McCarthy 2008), we see actors in ordinary organisations lowering their sights too, in terms of the ways in which they imagine that data mining might serve their purposes. There are no grand desires to know the social in its completeness in these examples. They are thus more modest in character and consequence than the more spectacular forms of data mining discussed in the book's introduction. As such, they point to the need to differentiate types of data mining. Indeed, we might say that in these examples, data mining is used to serve the public good.

Of course, concepts like serving the public good and engaging publics can be problematic. They can play a role in disciplining subjects into what counts as acceptable engagement. Cruikshank (2000) argues that actions which aim to engage or empower are proposed as solutions to social problems, but in fact, the 'will to empower' which is behind these propositions is 'less a solution to political problems than a strategy' for regulating citizens (2000, p. 1). Participation and engagement are posed as solutions to a perceived lack—of the right kinds of participation or engagement, she argues. I do not fully agree with Cruikshank's argument: I argue later in the book, specifically in relation to the examples of data activism and social research that I discuss in Chapter 8, that it is empirically inaccurate (Banks 2007 again) to cast such practices, and the public sector data mining experiments I discuss here, in only negative terms, as Cruikshank's argument suggests we might. That said, her argument speaks to the uneasy feeling I had when I was invited to participate in research about how digital technologies might engage publics: I wondered what it was that publics were assumed to be lacking and what was wrong with their existing forms of engagement that required academics to step in and point them in the direction of more appropriate, digitally enabled forms. So despite what I see as the limitations of Cruikshank's argument, it does highlight some of the problems with notions and practices of engagement and empowerment. My argument is that whether public sector uses of social media data mining concern us depends in part on how we view the public sector organisations using these methods and their aims and intentions. We could see them as extensions of the surveillance machine and its will to discipline subjects, or as local bodies doing what they can to serve their publics and enhance democracy. I suggest that both positions are useful in relation to social media data mining: we need to be aware of possibilities and alert to dangers. Having done the first of these—that is, discussed some of the possibilities that social media analytics presents to public bodies—in the next section I highlight some of the problems that we encountered in our action research, particularly in the ways in which data mining constitutes publics. Together, these difficulties, the desire for numbers that doing data mining provokes, and other organisational and contextual pressures, made it difficult for us to reflect critically with our partners about the making and shaping of data as we had hoped we would.

CONSTITUTING PUBLICS

How Keywords Constitute Publics

As part of our action research project, we suggested that our partners identify topics of current concern on which to concentrate our experiments, as we felt that this might be a focused way of exploring data mining methods. The museums group chose to focus on mining social media conversation about (a) a photography exhibition and (b) a set of online learning resources they had developed for use in museum education. One council chose to focus on finding out what was being said on social media about (a) council budget cuts and (b) their new health and well-being strategy. The other city council was interested in investigating social media talk about the Tour de France, which would visit the region soon. This partner's second chosen topic was a new market which was due to open up in the city centre. In many ways, the choice of topics was not of great importance, as our intention was experimental; with these topics we aimed to give our experiments some focus.

We asked our partners to identify 20 keywords for each topic, words that they imagined their publics might use when commenting on them. These would serve as starting points for our investigations. Keyword selection is an important component of data mining: an image of the phenomenon under investigation is created through the keywords that are used to describe it and this image is reflected in the results generated. In this sense, keyword selection is an aspect of data mining that brings the subject of enquiry into being. Many of the keywords suggested by our partners in relation to their chosen topics produced no search results, as these terms were not used by publics in online conversation. Searches in DataSift using the keyword 'telehealth' (in relation to health and well-being) returned no results, as did specific phrases such as 'fortnightly bin collections' in relation to council budget cuts. Other keywords, such as 'Moor', the name of the new city centre market, produced results that were too broad to be useful. To ensure that some data was found, the research team used keywords provided by partners as starting points for identifying other, more widely used terms. For example, in relation to the photography exhibition, we used the Flickr search Application Programming Interface (API) to find search terms used by Flickr members to tag images that were also tagged with the keywords supplied by our partner, and then used these newly found terms to search for conversation about the exhibition. The keywords we identified were more successful in generating data than those provided by our partners. We intervened in a way not initially intended, utilising our (at the time not extensive) expertise to overcome the limitations in our partners' knowledge, to ensure that some data was found and so enable our partners to reflect on its potential usefulness. We intervened so that our partners got what they wanted from their collaboration with us, and so that data mining methods delivered what we all expected them to: data, or numbers. This is one example of the research team acting as intermediaries, prioritising intervention over enquiry, action over research, playing a role in constituting the resulting publics and responding to the desire for numbers which was a feature of our research.

How Expertise (Or Its Absence) Constitutes Publics

The discussion above shows that expertise in keyword selection is necessary in order to be able to use data mining methods effectively. Social media data mining requires expertise of many kinds, not just in relation to knowing how publics talk about relevant issues. As noted elsewhere, Manovich (2011) argues that there are three ways of relating to data: creating them, collecting them and analysing them. In Manovich's assessment, only this latter process requires expertise. boyd and Crawford (2012) argue that who has big data expertise determines both who controls the data mining process and the 'knowledge' about the social world that results, knowledge which, as Law and others (2011) suggest, constitutes the social world. But access to data expertise is uneven and this leads to new digital divides, or the reproduction of old ones. Our research aimed to confront the danger of a digital data divide by experimenting with free tools, but the use of these tools did not circumvent the danger of an *expertise*-based digital divide. On the contrary, we found, it served only to highlight it. For, in contrast to Manovich's suggestion that only analysts require expertise, it was apparent in our research that expertise is also required in order to generate (or 'collect') data, as highlighted in the discussion of keyword selection above. Expertise is needed to use data collection tools like DataSift, not only to understand their interfaces, but also the fields of data held in records returned by the APIs of social media platforms. Of course, data analysis also requires expertise. As noted above, some expertise is needed just to find NodeXL, as it is an Excel template file, not a standalone programme. Once opened, understanding of the terminology used is needed. Then, users need to know how to make sense of search results (Figure 4.3 shows an example), how to manipulate results so they are visualised in meaningful ways, or how to export results into other tools, like Gephi, to produce visualisations.

This need for expertise was readily acknowledged by representatives from our partner organisations. One said 'you need to know the software inside out. You need to understand how to get into the data using that software' (Council Communications Officer). Even the commercial tools that we trialled were experienced as difficult to use by our workshop

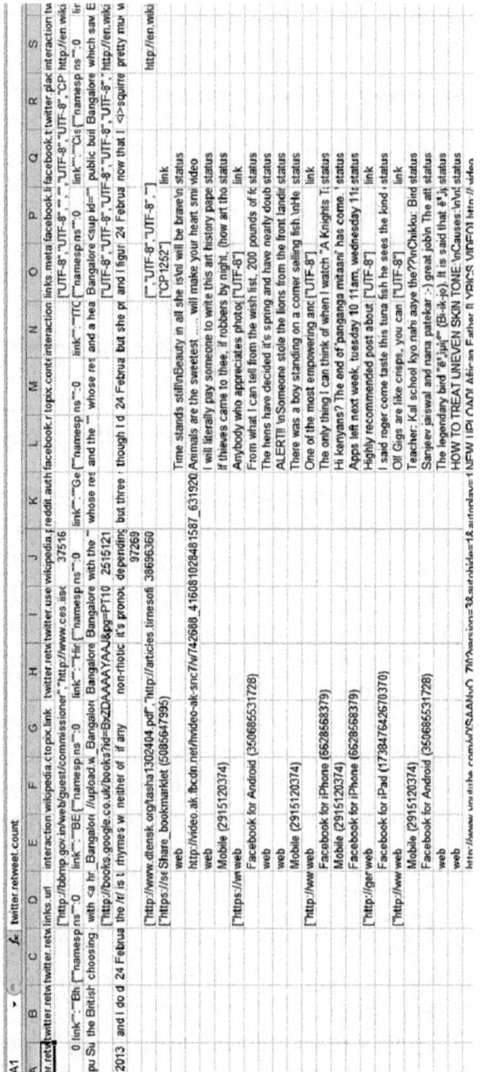

Figure 4.3   The results of a search as displayed in NodeXL

participants, perhaps somewhat surprisingly, given efforts made to produce usable graphical interfaces, a sample of which is produced in Figure 4.4. However, they still require certain kinds of expertise, for example in writing Boolean searches, and so require investments of both time (to gain expertise) *and* money (to purchase a licence).

It might seem obvious that expertise is needed to use data mining tools like NodeXL, but the complexity of such tools, which are public in the sense that they are freely available, limits their usability by public organisations. As I have shown, to address this barrier, the research team intervened in various ways. We changed keywords through processes of iterative search and we brought new tools into the project toolset in the hope that they would be easier for our partners to use, more successful in finding data and so would fulfil our partners' desire for numbers. In doing this, we enabled partners to find data, as more data was found with tools introduced later in the research. Representatives of the commercial tools we trialled intervened in similar ways. During one demonstration, the Account Director running the demo used keywords identified by our museum partner for the photography exhibition in combination with the names of towns and cities in the broad geographical area. She did this, she said, because someone writing online about a photography exhibition nearby might be interested in the exhibition our museum partner wanted to promote. This person's actions, like ours, shaped the data that was generated. We all intervened in these ways because of perceived barriers to tool use and because all of us, the research team, our partners and commercial tool representatives, engaged in action research with data mining methods with the expectation that data would be produced and that the desire for numbers would be fulfilled. Applying our expertise to the data mining process, we played a role in constituting the publics that resulted.

Working Around Data Non-abundance to Constitute Publics

One of the most widespread assumptions about digital data relates to its volume; allegedly, digital data are in abundant supply (Anderson 2008). The major social media platforms themselves affirm this abundance, as noted in Chapter 2. Such assertions produce expectations that data mined through related methods will be voluminous and easily found and, in this way, they produce a desire for numbers, in turn a key feature of datafication. However, there is great disparity between the data that are available

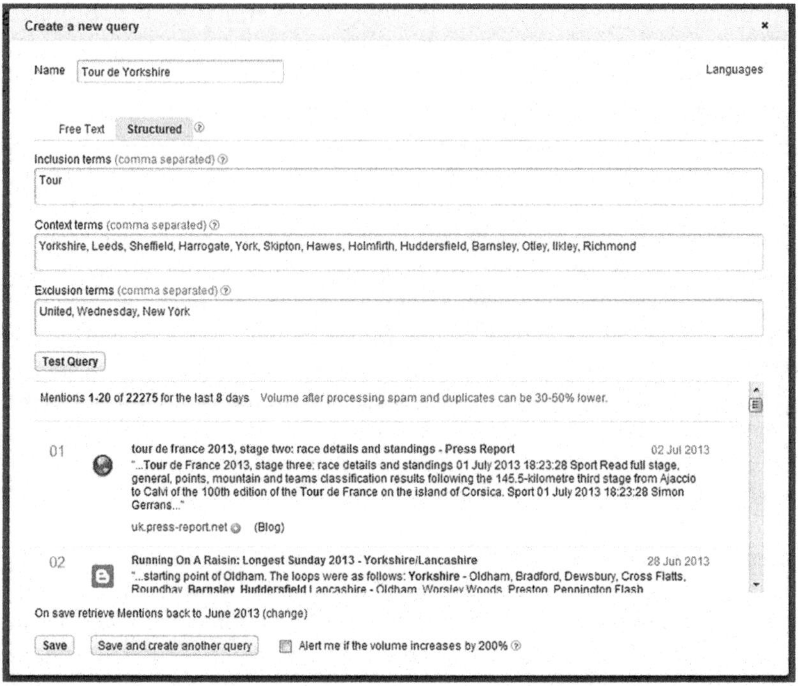

Figure 4.4   A sample interface from one of the commercial tools

in relation to large-scale global topics and the data which relate to small-scale local topics of the kind that our partners were interested in exploring. It is not the case that vast quantities of data are always there to be analysed, and finding relevant data in this mass can be challenging.

When setting out to find data about the topics chosen by our partners, one of our first steps was to explore online sources manually, through keyword searches for relevant conversations and by looking for key platforms used to discuss chosen topics. We carried out web searches using lists of known individuals and groups provided by our partners and compiled the URLs of websites to which they contributed commentary and opinion. Our methods of manual investigation allowed us to identify important conversation platforms, such as local city forums and comments sections of local newspapers' websites. This process was not time efficient, nor did it capture a large sample of content. But it served the important purpose of showing us that some relevant data were out there, and where. In contrast, as we started to use automated tools to search for data on a larger scale, the platforms that we had identified did not feature in results, and limited data were generated. Data shortage was sometimes because of lack of expertise in keyword selection, discussed above, but even when efforts were made to improve keywords, results improved only slightly.

On the whole, being local organisations, our partners were interested in finding local conversations and local 'influencers' with whom to engage. However, very little social data contains accurate location information. For this to be available, a location aware device with location services turned on is required, and users need to have agreed to their location being shared. Alternatively, location data can be derived from social media platforms themselves but, again, such information is not widely available (Graham et al. 2013). Utilising geographic filters to limit harvested data to that which is generated in target areas diminished an already small pool of data and excluded relevant contributors, such as local people writing comments on newspaper websites, forums, blogs and, in most cases, Facebook and Twitter too, who were not sharing their geographical location in any way. So with this approach, while the relevance of the data can be ensured, it is hard to derive general conclusions from them. As a Communications Officer from one of the councils put it:

> So, here, this is a really small number of people to draw any kind of conclusions from really. It's not representative of an audience in (the city). It is tiny really but like all data you kind of take it with a pinch of salt and pull from it the bits that you think are relevant and have got the most authority.

Tools are designed to source and analyse data in different ways, and these choices shape the resulting data. DataSift, for example, does not search the local platforms where our partners might find relevant data, like the city-based forums or regional newspapers' comments sections which we identified in our manual search. The tool user can make only limited decisions about where to search, from a finite list of the platforms which the tool developer considers to be relevant and has included in the tool. On DataSift, the user chooses which social media to pay to access, from a limited list of major platforms. Our ability to intervene in relation to data non-abundance was more limited than in the previous two examples because we could not influence how the tools operate, but we did what we could. We pointed out the absence of city-based forums among platforms searched by the commercial tools to the Account Director at one of the companies, who subsequently added these forums to the tool. Once again we intervened to enable our partners' desire for numbers to be met and, in so doing, we contributed to shaping the publics that emerged.

CONCLUSION: WHAT SHOULD CONCERN US ABOUT PUBLIC SECTOR SOCIAL MEDIA DATA MINING?

Through this action research, we explored how social media data mining methods are or might be used by actors within public sector contexts and how these methods shape the publics that they aim to understand. We experimented with these methods to examine whether groups that do not have the economic means to pay for commercial data services and want to use available data might access it. Because of our interest in addressing these different questions, in our study, we attempted to remain open to both the potential and problems of data mining methods, exploring empirically the assertions of data critics and advocates alike. But data mining and action research methods combined to produce the desire for numbers that I have discussed throughout this chapter, and this desire limited our success in balancing both, as our commitment to ensuring our partners were able to access data meant that we sometimes became the advocates that methods need, advocating for the methods about which we wanted to think critically. While different combinations of researchers, tools, partners and contexts might have produced different results, this tendency to privilege action over inquiry is not unusual in action research, given its commitment to social change, as Hammersley (2002) suggests. I argue that because data mining is motivated by a desire to produce results, data mining methods themselves produce the expectation that data will be found, that results will be produced, and that actions might be taken. They produce a desire for numbers, and this desire limited our capacity to think critically with our partners.

We encountered this desire for numbers throughout our research. For example, some participants reported that they were required to report the results of analytics exercises 'up' to managers and funders but that there was no discernible action taken as a result—the 'data gathered' box was ticked, the desire for numbers was fulfilled, and data were filed away. We noted that less data mining was taking place in partner organisations than had appeared to be the case in earlier interviews—this could be seen as reflecting a belief among interviewees that they *should* be doing data mining and producing numerical evidence of their actions. Conditions of austerity, datafication and its rhetoric, and 'mimetic isomorphism' (Granovetter 1985) combine to produce a desire for numbers.

In such circumstances, it is not easy to circumvent the danger of new, data-driven digital divides. The expertise needed to use tools proved challenging to our partners: they found it hard to identify relevant search terms and to find and gather data. Issues of resourcing, commitment, skill, access and data availability meant that not much social media data mining could happen in the public sector organisations. At the time of our project, all three partner organisations were in the process of drafting new social media policies for their staff and their participation in our project fed into this. Partners tell us that the reports we shared with them, the guide to tools we created and their attendance at our end-of-project workshop informed the drafting of their social media policies. In one partner organisation, the communications team held workshops to introduce staff to a range of social media data mining methods. In this respect, organisational knowledge of social media data mining which our research enabled was useful, according to our key contacts, and the action we took produced results. We appear to have addressed the danger of data-driven divides in a very modest way. But we are ambivalent about having this impact, not only because our research contributed to produce a desire for numbers within partner organisations, but also because this was not accompanied by the critical reflection for which we had hoped. Our experience raises the question of whether it is possible to do critical research *with*, not just *about*, data mining methods. I return to this point in Chapter 8 in my discussion of social researchers' experiments with social media data mining (for example Marres and Gerlitz 2015), which aim to work within this tension between exploring potential and recognising problems with data mining methods, to simultaneously use and critique them.

Our experiments made visible to us, the research team, some of the concrete ways in which elements of the data mining process construct the objects it aims to uncover. The selection of keywords, the application of different kinds of expertise and strategies deployed to work around data absence contribute to the production of particular results over others. There are many more examples like these, which show how different elements of data mining play a role in constituting the things they aim to represent. But critical discussion of these issues did not extend beyond the academic research team. Some partners expressed reservations about what they saw as the limitations of mined data: data were recognised as too imprecise to suggest specific actions, or too unrepresentative to be seen as reflecting the views of publics. An interviewee in one of the commercial insights companies which I discuss in the next chapter told me that in 'public sector organisations, there's almost a reticence, are we allowed to have this type of information?' This points to a degree of reflection about the ethics of data mining among public sector participants. But while some concerns were expressed about the appropriateness of intervening in social conversations through data mining in interviews carried out prior to the action research, during it, there was no discussion of this, of how citizens might feel about having their data mined or of whether organisations like councils and museums should be transparent about their data mining activities. There was no discussion, either, of whether social media data should be considered public or private. I argue that the conditions outlined throughout this chapter produced these absences, so at issue here is not a lack of individual actors' ethics, but rather a presence of over-determining conditions—structures of datafication produce an overwhelming desire for numbers. There was not much room for agency in relation to the issues discussed here.

Under such circumstances, it is not surprising that the publics produced through our data mining experiments remained thin and limited. In the case of our modest experiments, we were a long way off producing the fuller calculated publics that I imagined early in the chapter. We were dealing with extremely small datasets, often in their 10s, from which it was not possible to produce full, reflective and knowing publics. Local councils' and museums' uses of data mining—to understand how their publics are networked to each other, to identify influential individuals or groups to whom to promote events, or to identify the most beneficial social media channels to use to communicate with publics—do not invite publics to know themselves reflexively. But despite these limitations—in relation to critical and reflective conversation with partners and to the types of publics that our small-scale explorations produced—public sector social media data mining can serve positive ends. Organisations' aims of understanding publics in order to engage and involve them, and of including groups formerly excluded, mean that the consequences of these forms of data mining might be for the (public) good. The will to engage and empower can be seen as a strategy for regulating citizens, but it is not only this. Public sector uses of social data mining are not without problems, and they are not simple examples of the subversion of data power for the social good. However, it is unhelpful to cast public sector social media data mining *only* in a negative light. In our research, the types of publics produced by and the scope for reflexivity about data mining were limited but, nonetheless, public sector organisations attempt to do good with data in their uses of these methods The next chapter considers whether the same is true of data mining undertaken in commercial social media insights companies.

CHAPTER 5

Commercial Mediations of Social Media Data

INTRODUCTION

To undertake social media data mining, public sector organisations like those discussed in the last chapter sometimes engage the services of commercial social media insights companies. As indicated in Chapter 2, such intermediary companies, which analyse social media activity on behalf of paying clients, play an important role in making social media data mining ordinary. They offer a broad range of services to a broad range of customers, such as identifying how many people are talking about a particular topic or brand across social media platforms and what they are saying, finding key influencers and the location of key conversations, and extracting demographic information about people engaged in social media conversations. They often promise, directly or indirectly, that such insights will help clients to increase efficiency, effectiveness and profits.

To date, very little research has been carried out into the work undertaken by these emerging companies, despite their significant contribution to social media data mining's growth in scale and scope. So it is important to turn our attention to the role of these players in the 'eco-system of connective media', as van Dijck (2013a) calls it. Two writers who have directed critical attention to the sector are Hearn (2010) and Andrejevic (2011), whose work was discussed in Chapter 3. Hearn (2010) has argued that some forms of social media data mining monetise feeling and friendship and so operate as mechanisms of capitalist value extraction, while Andrejevic (2011) is critical of the role played by data mining, especially sentiment analysis, in the prediction and control of affect. Hearn and Andrejevic's critical interventions do the important job of highlighting some of the troubling consequences of data mining that I have discussed in earlier chapters but, nonetheless, there remains an absence of empirical detail about the sector, how it operates and how people working within it think about some of the core issues discussed in this book. Such studies are needed in order to develop understanding of the work of social media insights companies and the extent to which their activities should concern us. This chapter represents an effort to fill that gap.

I start the chapter with a brief sketch of this intermediary sector. Mapping cultural industries is difficult, and this one is especially so, as it emerges from diverse fields. The social media data mining services that interest me are offered by a range of companies, not all easily identified as belonging to one narrow 'sector'. So the sketch is incomplete, but still it starts to build a picture of the field. Without doubt, this fast moving sector will have moved on by the time this book is published, so the outline provided here is best understood as a brief history of an important moment in the sector's development, as social media data mining became ordinary. To give one example of such changes, at the time I carried out my interviews—more on these below—some companies used the phrase 'social media monitoring' to describe what they did, while others used 'social (media) insights' or 'intelligence'. By the time of writing this chapter in 2015, the term 'social media monitoring' is much less widely used, perhaps because of the surveillant connotations of the word 'monitoring'. I use these and other terms ('insights', 'intelligence', 'monitoring', 'analytics', 'data mining') interchangeably here to reflect company language, conscious that new terms may come into usage in the time between writing and publishing the book.

The chapter proceeds to address the question of what, if anything, should concern us about the work undertaken by these companies by focusing on the norms, ethics and values of workers in this sector, and how these shape the work that gets done. I do this, first, by discussing the concerns of the workers themselves. These often relate to the quality and accuracy of social data. Interviewees' concerns about the lack of both in social data shows some awareness of methodological limitations of social media data mining discussed in Chapter 3, something that was generally absent among the public sector workers who were just beginning to experiment with data mining discussed in Chapter 4. Some interviewees from the commercial companies expressed frustration that their concerns were not shared by their clients. As in the previous chapter, I see a prevalent desire for numbers, emerging from conditions of datafication and more historical patterns of faith in the quantitative, as serving to limit the possibility of opening up a space for conversation about the ways in which data are made and shaped.

The second half of the chapter focuses on ethical issues. I discuss responses to questions put to interviewees about surveillance, privacy-invasion and transparency in data mining, concerns I outlined in Chapter 3. I invited interviewees' responses to related criticisms in order to explore the role that ethics play in their workplace decision-making. What emerges is an ethically complex picture, in which many decisions are made on both ethical and economic grounds. This is not an unethical field: some interviewees consider some practices to be ethically off-limits, but ethical lines are not drawn in the same place for all workers. To make sense of this ethical diversity, I draw on the concept of a 'moral economy', which has been used in cultural industries research (for example, by Banks (2007), Hesmondhalgh and Baker (2010), and in my own work (Kennedy 2011)). In this work, the proposal that economic decisions, behaviours and institutions 'depend on and influence moral/ethical sentiments, norms and behaviours and have ethical implications' (Sayer 2004, p. 2) is mobilised to propose understanding cultural labour not only as resulting in economic value for those who own its outputs, but also as a process which involves a series of judgements based on the values of workers themselves. This assertion that value and values come together in cultural labour applies to social media data mining just as much as any other form of cultural work, because data miners' values play a role in the ways in which they carry out their work. But because of the prevalent desire for numbers discussed in the previous chapter and in this one, the concrete consequences of this ethical thinking are rather limited, as I show below.

The chapter draws on 14 interviews I undertook in 2012 and 2013 to explore these and other questions, in companies with headquarters based in the UK (8), the US (1), South Africa (1), Norway (1) and Spain (3).[1] Of these companies, five can be described as social insights firms, and four are digital marketing agencies offering a range of services, including social media insights. Two are media monitoring companies which offer social media as well as traditional media monitoring services, two are digital reputation companies which incorporate social media data mining into their processes, and one is a sentiment analysis company which does not focus solely on social media. Table 5.1 provides further detail about

---

[1] Thirty-eight companies were identified and approached on the basis of online searches and offline contacts, of which 15 agreed to interview, but, in the case of one company, it was not possible to find a suitable time to carry out the interview.

Table 5.1 Social media insights interviewee and company details (anonymised)

Company name	Company details	Interviewee name	Interviewee details
Social (media) insights companies			
Witness	Social insights company, 300 employees, UK headquarters, offices in six countries.	Amanda, Account Manager	Background in cultural studies and in working in SEO (search engine optimisation).
ProductWatcher	South Africa-based online insight software company, 17 staff, including CEO, marketing, sales, insights specialists and development team.	Margie, Sales and Marketing Director	Marketing background, in large companies and PR agencies.
24-7Social	Corporate social insights and analytics, offices in London and New York, global staff, 15 core staff as well as additional analysts and writers.	Lawrence, COO (Chief Operating Officer) and co-founder	Background in corporate/financial PR, previously ran a technology and telecoms PR agency.
Fusage	UK-based, describes itself as 'monitoring social intelligence'. Recently named as one of top 11 listening platforms by Forrester Research.	Alexander, Analyst	Market research and media planning/buying background (for example, campaign evaluation using metric modelling)
Detector	Small, Spanish-based online intelligence platform, set up by the interviewee.	Patricia, CEO	Studied business management and marketing, worked in business marketing, studied internet marketing after some years of working.
Digital marketing agencies			
Paste	UK-based, award-winning, background in paid and organic search, 18 senior staff, up to 100 other staff.	Gail, Head of Social Media	Communications background, including interpreting for deaf people, online PR, social media.
BrandHook	UK-based, background as an SEO company.	Robert, Head of Social Media	Studied politics and sociology, has been working in social media throughout his short career and since its origins, in agencies and large companies, a social media early adopter.
Octopus	UK-based, background in paid search and PPC (pay per click), branched out into SEO and social media.	Isabel, SEO and Social Manager	Studied fashion and PR, worked as web and social media editor for Girl Guiding, active volunteer.

Discern	Spanish digital strategy consultancy, offering three services: strategy, human resources and training, and client loyalty.	Susan, Head of Social Media Marketing	Studied advertising, PR, business management, set up own internet skills training company in 1995, internet early adopter, worked in online reputation management, media monitoring, web development, then set up this company.
Media monitoring firms			
BlueSky	Traditional media monitoring company which has branched out to offer social media monitoring services.	Isla, Account Manager	Studied international relations, including role of social media in Western perceptions of conflict, which influenced her getting this job.
Claimr	Norwegian, traditional/manual media analysis and content analysis company. Uses manual techniques.	Bernhardt, Analyst	PhD in media studies, went on to work for Google after interview.
Digital reputation services			
Checker	UK-based, set up by interviewee in response to emerging issues re online reputation. Offer verification services to protect businesses from defamation and customers from fraud.	Daniel, CEO and founder	Background in corporate communications and PR.
Rely	Global company, 40 staff, headquarters in San Francisco, previously did pure social insights, shifted to incorporate these into reputation service. Now offers social media influence measurement.	Gareth, CEO	Worked in marketing and engineering (mobile advertising, location-based services), worked for competitor.
Sentiment analysis platform			
SentiCheck	Tool for understanding language use, predominantly used in business management, or 'people development within organisations'. Based in Spain.	Graciela, founder and developer	Research background in psycholinguistics and neuroscience, developed SentiCheck tool in collaboration with academic computer scientists.

the companies. As this categorisation of companies suggests, commercial social media data mining has its origins in a range of persuasive communication practices. Joseph Turow (1997, 2008, 2012) has written extensively about the pre-history of contemporary forms of commercial data mining, locating their origins in the advertising industries, which have long used similar techniques to monitor, measure and understand target audience psychographics (such as values, attitudes, interests, lifestyles). Similarly, Arvidsson (2011) argues that social media monitoring practices have historical precedents, not only in the use of pyschographic variables in advertising but also in the rise of value attached to ephemeral phenomena like 'the brand'. These authors and others (such as McStay 2009) identify continuities between social media monitoring and other forms of persuasive communication like advertising, marketing and PR. While this pre-history is important—interviewees identified their own origins in these sectors too—this is not a historical book, so I focus my discussion on contemporary practices and what my respondents said about them in the interviews I carried out. These aimed to establish the work biographies of the interviewees, explore the work undertaken by the companies, identify ethical codes or codes of practice to which they adhered, and to put to interviewees some criticisms of social media data mining and allow them to reflect on/identify their own (ethical or not) positions in relation to these concerns. The chapter also draws on online research and textual analysis of the websites of social media data mining companies, as well as field notes from two funded internships in digital marketing agencies undertaken in the summer of 2012. Research assistants who helped me with these aspects of my research are: Christopher Birchall, Patrick McKeown, Cristina Miguel, Matthew Murphy and Stuart Russell.

The Practices of Intermediary Insights Companies and the Concerns of Workers

Who Companies and Interviewees Are

As noted in Chapter 2, according to Francesco D'Orazio, Chief Innovation Officer at Face, which runs the social insights platform Pulsar, at the end of 2013, there were more than 480 available social intelligence platforms (D'Orazio 2013). Social media data mining services are expanding around the world, especially in the global North, he suggests. And yet the sector is still in its infancy, as Soenke Lorenzen of Greenpeace suggested in a keynote lecture at the Digital Methods Initiative Winter School in Amsterdam (2015). As such, it is fragmented and vendors come in all shapes and sizes, he continued, although major players dominate. While D'Orazio does not specify which social insights companies are included in this figure, some online reviews attempt to name the most significant players in lists of top 10s, 20s and 50s. None, however, provide a comprehensive list. The website Social Media Biz (2011) produced a comparison of the top 20 social media monitoring platforms in January 2011, and around the same time, independent technology and market research company Forrester Research started to produce reports which aimed to identify and assess key players (for example Hofer-Shall et al. 2012). More recently, Social Media Today (2013) produced a similar list, this time with a top 50 rather than a top 20, reflecting an expansion in the numbers of companies operating in this sphere. Also in 2013, UK-based social listening company Brandwatch produced a list of the top 10 free social media monitoring tools (Mindruta 2013), and TweakYourBiz.com also produced a top 10, this time not restricted to free tools (O'Connor 2013). The most recently available lists at the time of writing include a post from 2014 on Social Media Biz identifying the top 10 tools for monitoring Twitter and other social media platforms (Totka 2014) and a Forrester Wave report from the same year on 'The 11 (enterprise listening) providers that matter most and how they stack up' (Smith 2014). Alongside platforms offering free services like HootSuite, TweetReach, Klout and Social Mention, many of which were discussed in Chapter 2, commercial companies dominating the English-speaking market which are widely acknowledged as key players at the time of writing include: Sysomos, Radian6 (now part of SalesForce Cloud), Alterian SM2, Lithium and Attensity360, most of which are US-based, and Brandwatch and newcomer Synthesio, which have their headquarters the UK.

One factor that makes it difficult to map social media insights companies is that this is not a boundaried field. Companies emerge from, relate to and offer the same services as a number of sectors, including marketing, market research, search, PR and web analytics. As well as specialist social media insights companies, digital marketing agencies, search and PPC (pay per click) companies (such as BlueClaw and StickyEyes in the UK), market research firms (like Nielsen), web analytics firms (including Google, through Google Analytics and Google Alerts), digital reputation management companies (for example Kred or KwikChex) and software companies (like Cision) increasingly offer social media insights, making it hard to assess the number and range of companies operating in this field. Table 5.1 provides brief details about interviewees and the companies in which they worked (all anonymised here), as well as giving an indication of the range of their 'origin stories'.

What Companies and Interviewees Do

Some companies (called social media insights companies here) describe their services as data provision or data reporting, whereas others (often the digital marketing agencies) incorporate social media data mining into fuller digital marketing services. Interviewees from this latter group of companies described similar processes to me; I provide a sketch of one such narrative here to paint a picture of how they operate, the concerns they have and the issues they encounter. The Head of Social Media at Paste, whom I call Gail, described her background as being in communications, not only because she had worked for online PR agencies but also because both of her parents were profoundly deaf and she had spent much of life as an interpreter for them and other deaf people. Her company describes itself as a digital marketing agency with a background in paid and organic search. The process of working with clients starts with an audit, an in-depth piece of work which analyses how clients' customers are using social media ('So if they're all over Pinterest, we're all over Pinterest,' said Gail), the types of marketing tactics to which they are open, such as online conversations, price promotions, and the brand's own or competitors' social media activity. The audit also includes an evaluation of key influencers in relevant domains, with whom clients may wish to engage.

This can be the end of a client's involvement with Paste; they can choose to produce their own marketing plans on the basis of the audit, or they can continue to work with Paste on the development, implementation and evaluation of a digital marketing strategy. Such strategies are closely tied to the client's business and objectives and, as Gail put it, they 'aim towards making more money'. Clients are keen to identify the precise return on investment (ROI) of social media activity, something that Gail described as her 'worst enemy, I have nightmares about it every night basically', because identifying a precise ROI in relation to social media marketing and data mining is difficult. It is hard to establish a causal relationship between client social media activity and customer purchase habits (did a customer purchase a product because of a social media campaign, or a word-of-mouth recommendation?). But despite this difficulty, Paste attempts to meet clients' desires for precise data, through statistics in case studies on its website, such as numbers of viewers reached or page views, numbers of links from influential bloggers and percentage increase in year-on-year socially attributed revenue, this latter sounding very much like ROI.

Paste offers training on how to implement the digital marketing strategies that it produces, in the hope that clients will manage their own, but this rarely happens, and instead, clients ask Paste to implement it on their behalf because they 'just can't deal with this headache'. This concerns Gail. 'I always set out in my business plan to say we shouldn't be taking that on, we should let our clients look after their own voice', she said, because the only people who can talk in an authentic voice about an organisation is people within that organisation, she feels. For her and other interviewees, authenticity is an issue. But despite her concerns, most clients are content to have Paste manage their campaigns in a voice which she describes as 'good enough for most of our clients'. This means that other major services offered are campaign implementation and evaluation, which involve the use of a range of in-house, proprietary and free tools to measure performance, reach and returns.

Social media insights companies which emerged from market research offer different services and highlight different issues. Alexander, Account Manager at Fusage, which describes itself as 'monitoring social intelligence,' had background in market research, where what he described as 'hardcore data analysis' was commonly used to evaluate the effectiveness of advertising campaigns. The core issue that he identified in relation to his current role was the quality of the data that is available on social media. He stated that when his company started out in social media intelligence, their data miners were sceptical about what could be done with what they saw as the inconsistent and unstructured data that social media platforms produce, which were considered to be not sufficiently dependable for analysis. To traditional data analysts, social media data can appear unstructured, messy, methodologically flawed and impossible to work with it: he described them as 'volatile data'. Talking about the reliability of metadata, he gave the example of research carried out for a skin cream company, which suggested that the majority of people talking about its products on social media were young men, not the company's known audience of mature women. He concluded: 'So it's just basically whoever registers their details is your demographic audience according to social media. [...] No matter how good your Boolean is, you're always going to get stuff that shouldn't really be there.' Patricia at the social insights company Detector confirmed this, pointing out that relevant data is easily mixed with irrelevant data for companies with names featuring common words, like Orange and Apple.

Evidence of these concerns about quality and accuracy of data can be seen elsewhere. In a newspaper article in 2009, Fowler and de Avila pointed out that there was a 'positivism problem' in social data, as the average rating for all things reviewed, from printer paper to dog food, was 4.3 out of 5 (though this may have changed by the time this book is published). They quoted Ed Keller of market research group Keller Fay who says that 'there is an urban myth that people are far more likely to express negatives than positives', whereas the opposite is true. In their surveys, Keller Fay found that around 65% of reviews were positive, whereas only 8% were negative. Likewise, Fowler and de Avila pointed out that some websites acknowledge that companies may be submitting reviews of their own products, and that negative reviews may be suppressed, further indication of this 'positivism problem'. In addition, one of my respondents, Daniel, who runs a reputation management company, suggested that more than 10% of online reviews may be falsely negative and so also contribute to the inaccuracy of social data. Insights workers' concerns about the quality and accuracy of the data with which they work suggest an awareness that social media data do not simply and straightforwardly provide insights into social world. Nor are they just collected. Rather, they need to be acted upon, cleaned and ordered, to be useful and usable: when asked whether his company analyses data on behalf of clients, Lawrence of insights company 24-7Social, which he described as offering 'a straight reporting function', said 'the only analysis we do is taking the messy, unclear and ambiguous world of social media and putting it in a simple jargon-free way that the clients will understand'. Such observations, in turn, point to awareness among these social media data workers of some of the methodological issues and limitations of data mining discussed in Chapter 3.

If social media data are inaccurate, unstructured, volatile and difficult to work with, how do insights companies persuade potential clients to sign up for their services? One way is to assure potential clients that there is plenty of data out there to be mined (in contrast to what we found in our small-scale experiments with public sector organisations, discussed in the previous chapter). To do this, companies emphasise that the social media conversations that they track are real-time and high-volume. For example, one company's website states 'Sysomos lets you listen to millions of conversations about your brand and products in real-time' (Sysomos 2012). We saw this in the case of Paste above, who cite numbers reached as evidence of the value of investing in social media data mining. In one case, they give the precise figure of a reach of 21,996,323. Andrejevic (2011) argues that claims made about services offered in this sector refer not to what he calls 'referential accuracy' (that is, that the data can actually be taken to represent what they are assumed to represent, or that they come from target demographic populations), but rather to the volume and real-time-ness of tracked data. Size and immediacy make up for the roughness of the data, he claims.

Most of my interviewees acknowledged that they often encounter irrelevant, limited or inaccurate demographic data and obstacles relating to what tools cannot do, and many suggested that clients do not mind that this results in 'rough' data. Talking about client expectations, Lawrence at 24-7Social said:

> Whether that data is accurate or not is irrelevant. They just want some numbers to put into a PowerPoint that they can show to their boss. If anyone asks, 'Are you keeping an eye on social media?' You can say, 'Yes, we're 36 this week.' And it is a very attractive solution.

In other words, it suffices to give clients a number. Clients desire numbers regardless of accuracy, in order to fulfil certain expectations, such as reporting to line managers, or providing evidence that they are monitoring social media appropriately. Gareth from Rely, one of the digital reputation companies, suggested that this formed part of a broader allure of numbers and statistics:

> When we talk about influence, a lot of clients are blindsided by the number on its own and think the highest number wins, when that's not really the case. So I spend part of my time doing education around what you do with the numbers and what to look for.

Some clients, it would seem, are not concerned that data is clean or 'referentially accurate'. Sometimes they are drawn in by the allure of numbers and just want numbers. When it comes to financial investment and returns, they demand precise numbers, but when it comes to social data, imprecision is acceptable, as long as the desire for numbers is fulfilled.

Those of us concerned about methodological problems with social media data mining might share the concerns expressed by the insights workers cited here. Like them, we might also want conclusions drawn and actions taken as a result of social media data mining to be based on accurate, clean and robust data. The fact that referentially inaccurate datasets might be used as a basis for decision-making that affects our lives might concern us. Interviewees demonstrate some understanding of Bowker's assertion that 'raw data is an oxymoron' and seem to acknowledge that data are generated in conditions shaped by human decisions, interpretations and filters. When asked if she felt that she was shaping the data that she outputs through the decisions that she makes, Isla at the media monitoring company BlueSky said: 'absolutely, 100%. We are manipulating keywords; we're manipulating the data effectively. That is what our service and what the keyword technology is designed to do. Of course I don't mean manipulative in a pejorative sense.' Similarly, Robert at the digital marketing agency BrandHook said: 'It's all about filtering and mining data and getting to exactly what the brand wants.' So it would seem that these commercial data practitioners do not propagate the notion that data can be raw. As we have seen, they recognise that data is messy, unstructured, not 'clean', robust or reliable, and that they need to 'act on' data in order for them to be useful and comprehensible.

In the interviews, a range of issues emerged—quality of data was mentioned across all the different types of companies, but authenticity was an issue just in those companies which offer full marketing services, and not in others which describe their function as data reporting. Some respondents were reflexive about the ways in which data are made, manipulated, inaccurate and volatile, others were not, for example the interviewee who suggested that that data can be found, cleaned and handed over. Social media data mining companies, practices and workers are not all the same; they differ, and therefore we need to differentiate when we talk about them. But they all operate in conditions of datafication, and these conditions produce a widespread desire for numbers, which, as seen in the previous chapter, can serve to suppress discussion about the conditions in which data are generated and the implications of these conditions for the reliability and accuracy of the numbers that data mining produces. These conditions need to be subjected to critical scrutiny, and cultural industries approaches offer a useful lens for doing this. Understanding data workers as producers, seeing data work as production and exploring the data production process can contribute to developing our critical understanding of social media data mining. Likewise, attending to the conditions of production can reveal the concerns that surface in commercial social media data mining work, as we have seen above. As noted in the introduction to this chapter, one approach adopted in cultural industries research, a moral economy approach, can open up a space to explore the role of ethics and values in production processes, which can also contribute to understanding how commercial social insights work gets done. I adopt such an approach in the next section.

SOCIAL MEDIA DATA MINING AS MORAL AND ECONOMIC PRACTICE

In an article entitled 'Moral economy', Andrew Sayer argues that economic decisions, behaviours and institutions 'depend on and influence moral/ethical sentiments, norms and behaviours and have ethical implications' (2004, p. 2). '(E)thical and moral valuation is always either present or latent' in economic behaviour, he writes (2004, p. 4). Here and elsewhere, he understands morals and ethics, terms he uses interchangeably, to mean 'norms (formal and informal), values and dispositions regarding behaviour that affects others, and they imply certain conceptions of the good' (2004, p. 3). Sayer is one of several British political theorists and philosophers interested in examining the relationships between ethics and markets. Like him, Russell Keat has also written extensively on the intersection of ethics, morality and markets, for example in 'Every economy is a moral economy' (2004) and 'Market economies as moral economies' (2011). In these papers, Keat argues that critical evaluation of market economies must include attending to ethical judgements about the goods and ills of production, consumption and exchange. Similarly, John O'Neill asserts that non-economic associations are central to economic life (1998, p. 76; see also Arvidsson and Peitersen 2009).

The writers discussed here make the case that labour not only results in value for those who own its outputs, but also needs to be understood as a process which involves a series of judgements based on the values of workers themselves. These ideas have been taken up in the work of leading cultural industries academics, such as Mark Banks (2007) and David Hesmondhalgh

(2010, 2014). In *The Politics of Cultural Work* (2007), Banks argues that we need to acknowledge that cultural workers are human subjects with psychological needs, which lend them to ethical, moral and social practices. Banks discusses a range of cultural production practices which result from such tendencies, such as ethical fashion houses, socially responsible design agencies, community arts organisations, public access media, not-for-profit design companies and art collectives. He concludes that it is an empirical misrepresentation of the efforts of many cultural workers to claim that they have *only* served capitalism through their work. David Hesmondhalgh takes up the concept of the moral economy in his book with Sarah Baker, *Creative Labour: media work in three cultural industries*, in which they develop a model of good and bad work, which includes engagement in the creation of products which '*promote aspects of the common good*' (2010, p. 36) as a feature of good work. In other talks and papers, Hesmondhalgh (2010, 2014) explores the ways in which different economic arrangements enhance or diminish the contribution of culture to modern societies, advocating a moral economy approach to addressing this question, which, he argues, helps to overcome more simplistic understandings of markets.

In previous research I carried out into the work of web designers, I used a moral economy approach to argue that web design is suffused with ethical inflections, and that many aspects of the work of web designers, such as their commitment to producing websites which are accessible to web users with disabilities, can be seen as ethical practices. In the case of web design, it was easy to see that 'ethical and moral valuation' was 'present or latent' (Sayer 2004, p. 4) in web designers' economic behaviour. It is not so easy to apply such a model to social media data mining, which, at first glance, seems much more ethically problematic than web design, for the reasons discussed in earlier chapters. However, if we argue that there are moral economies of cultural labour, then we should consider the role of ethics and values in *all* forms of cultural labour, including its more maligned forms, like social media data mining. Doing this might help us answer the question of what should concern us about social media data mining. In the following paragraphs, I show how, in commercial contexts, value and values merge, as values play a role in the ways in which social media data miners carry out their work. What emerges is an ethically complex picture, in which decisions are often both ethical and economic.

Before examining how ethical issues were discussed in the interviews I carried out with people working in commercial social insights companies, it is useful to reflect on respondents' backgrounds, as it shows that, like other cultural workers, some of them may have inclinations to moral, ethical and social practices (Banks 2007). One respondent, Daniel from the reputation management company Checker, had experience in what he describes as 'integrity-led' communications consultancy. Another, Amanda from the social insights company Witness, had worked in search engine optimisation (SEO), where the ethics of different techniques are hotly debated and where she had been a vocal critic of what are seen as unethical practices. Others were active volunteers, such as Isabel from Octopus, who in her short career had used her skills to support two campaigns, Beat Bullying and Child Exploitation Online. Others, such as Margie at ProductWatcher, give talks to young people, advise youth organisations on social media practices, or speak in universities, sharing their knowledge and experience with others. Several respondents had studied media, culture and communication. During my interview with Amanda, she said that she had recently been asking herself, in relation to the work that she does, 'How would Marx frame everything?' Not only is this a great anecdote, but it also indicates that workers' biographies are significant, for, as we shall see, the fact that critical media studies and ethical work practices play a part in the formation of today's army of social media marketers and data miners has some impact on the ways in which such work gets done. In the remainder of the chapter, I highlight several issues which involve 'ethical or moral valuation', including: accessing only public data; the ethical limits to what workers will do; being transparent; the question of who benefits from data mining; and attitudes to regulating the sector.

Accessing 'Public' Data

When asked if they have codes of practice regarding the data that they gather and analyse, all respondents said that they only mine publicly available data. If data is behind a firewall, or is not publicly available, then it is considered to be off-limits. 'If there's an indication that people don't want their website looked at, then we don't look', said Amanda at the social insights company Witness. This is guaranteed by the fact that the functionality to log-in to closed sites is not built in to her company's technology, although it could be. It is *technically* possible to breach systems' terms of service and 'walk under the fence', as Bernhardt at the media monitoring firm Claimr put it, for example by creating fake accounts, friending people and so accessing their private data. However, a more principled approach is usually chosen, guided by social media platforms' terms of service—that is, to do what platforms say that third party data miners can do. This decision is both moral and economic: interviewees think that 'under the fence' practices are wrong, but they also acknowledged that doing what is permitted by platforms is in their business interests. When asked if such decisions are ethically or economically informed, Bernhardt at Claimr said:

> Ethics is a part of it. But it's not that this was a big discussion overall. I think we just all sensed that this is not the thing to do, because we want to limit ourselves to the number of posts that are publicly available and not trying to get hold of more than that. We all felt that if our data that we had posted just for friends would show up in searches, that wouldn't be fine. There is an understanding that this is clearly off-limits.

Alexander, at the social insights company Fusage, acknowledged the importance of respecting privacy settings for legal reasons. He said: 'I don't think that it's that we think it's not right from a personal privacy point of view, it's that it's not right from a legal point of view. [...] I think it's much more a case of legal ramifications than personal liberty.' Thus approaches which appear principled or ethical can often be commercially or legally motivated.

Accessing only public data and respecting privacy settings is not as straightforward as it might seem because, as noted in Chapter 3, what is public and private is complex in social media environments. As observed in that chapter, boyd (2014) differentiates between being public and being *in* public. In the latter cases we might still expect a degree of privacy, and Nissenbaum (2009) emphasises the need for contextual integrity with regard to such expectations. So even companies which claim to protect privacy—such as the sentiment analysis company SentiCheck, where the interviewee Graciela outlined processes of ensuring confidentiality, anonymisation and quick deletion of data—cannot be sure they are dealing with public data in every sense of the term. I put this to my respondents. A number of them felt that responsibility for identifying whether data are public or private lies with platform users. For example, digital marketing agency Discern in Spain, where the interviewee was called Susan, lists among its clients a large multinational for whom the agency tracked its employees in order to identify 'what feeling they have' about the organisation. Susan argued that it is social media users' responsibility to understand how public their social media data are. She said: 'If I do a nude photo and upload it on Twitter ... whose fault is it? "Oh! I didn't read that this could be seen by everyone." ... So, you should have!' Another respondent, Amanda at Witness, continued with boyd's corridor metaphor to draw distinctions between speaking and publishing. She said: 'if you've published it online, then you've published it. You haven't said it, you've published it. You haven't said it in a corridor, you've written it on a corridor wall.'

Some respondents acknowledged that the distinction between public and private is ambiguous in social media and the status of their data is not always clear to users. Some noted that some social media data is public both from a systems point of view and because it is clearly directed at companies and appears to invite a response but, as Bernhardt stated, 'It's not crystal clear what's public and what is not.' Gail, the Head of Social Media at digital marketing agency Paste, recounted a tale from a project her agency had undertaken for a bingo company about customers' dream prizes which, as she put it, shows how people behave in intimate ways on social media platforms. She told me how one woman went into great detail about the conditions in which her son had recently died, and continued: 'I remember thinking I can't believe that anyone would want to share this information with a bingo website and tell them why they wanted to win money, I just can't fathom it.' Even Gareth from Rely, who had previously suggested that wanting privacy meant that you have something to hide (when I suggested that someone who did not want to be tracked might think 'I don't want you to monitor what I say', he replied: 'What have you got to hide?'), acknowledged this blurring of the public and private. He talked about how something shared privately can become public through someone else's re-sharing, using the example of Facebook CEO Mark Zuckerberg's sister, who had posted a Christmas family photograph that became public through such re-sharing (McCarthy 2012). So even though people working in related industries are aware that something shared privately on Facebook can end up public because someone else shares it, said Gareth, 'the Facebook privacy settings are so confusing that even the founder's sister can't get it right'. But he continued:

> in defence of the industry, we don't go out finding things that are private, we are only legally allowed to access things that are public, so at some point, a human being has leaked some information.

Lawrence at 24-7Social continued this defence of the industry. He defined his own company as 'ethically neutral', describing the services that his company offers as 'reporting to our clients what people are saying about them in a public forum' and he argued that, even if people are not fully aware that their social media data are public, 'if it is public it is therefore of interest to the person who it is about'. At the same time, and despite describing his work as ethically neutral, he demonstrated his own recognition of the ethical ambivalence of mining social data when he acknowledged:

> I am not sure anyone would welcome the monitoring that we do. I mean we don't, we would argue that we provide value back to them by ensuring that their company is better informed about what they are doing wrong and therefore would see the error of their ways. That would be the sort of positive message, but at the same time whether or not people want us to be monitoring what they are saying ... I still believe in the right of the company to know what is being said about them.

Drawing Ethical Boundaries

The above quote and others cited here demonstrate moral economy in action, as workers weigh up the ethical pros and cons of the decisions they make about the work that they do and, in turn, the profits that their companies make. Another way in which the moral character of economic decisions became visible in the interviews was in the ethical limits to what interviewees said they would do, or the ethical boundaries that they draw around their data mining practices. Bernhardt at the media monitoring company Claimr gave an example. He said that a public relations agency had asked his company to mine data about journalists, so that the agency could inform its clients of what individual journalists were writing about them. 'And we chose not do that', said Bernhardt, because 'this is not something we want to do. I don't think it would be in conflict legally, but we feel it's one step too far.' His company mines and analyses 'the general mood', he said, not individualised emotions or opinions. 'We're not going too much into the individual, but trying to catch the feeling', he said. As a result of 'trying to catch the feeling', as he put it, he felt that his company's practices were not harmful or intrusive to individuals. He continued:

> We do not place too much attention to the individual statements, but more the total picture of what is said in a given time period about a company. [...] We can identify persons if we want to, but we don't see the relevance of it.

Other respondents told similar stories. I had the following exchange with Lawrence at 24-7Social, which he had previously described as an 'ethically neutral' company, which is worth citing in full:

L   There are definitely dark areas where, Egypt and Tahrir Square, if someone asked you to monitor the tweets of people who were in Tahrir Square on behalf of the Egyptian government, that is a very different kettle of fish.

HK   Would you say no?

L   Yes, I don't think we would get involved. We have done stuff for Middle Eastern governments, so during the Arab Spring a couple of the governments wanted to know what the impact on tourism was of the instability. So that was more monitoring international comment about whether you would go on holiday to Egypt this year, kind of thing. You might be working for governments you don't agree with, but it's not for a sinister purpose, it's their Tourism Board. But from our point of view we wouldn't want (to monitor) individuals in a kind of revolutionary movement or anything like that, because it becomes very dark.

HK   So not totally ethically neutral?

L   Not totally ethically neutral.

HK   So there is a line that you draw?

L   Yes. As an organisation we are ethically neutral, as people we are not.

Here Lawrence highlights that workers in social insights companies bring their own ethical codes to bear on the work that they do for economic accumulation. These workers are human subjects with ethical, moral and social values. They have individual, ethical barometers—they draw lines around what they will and will not do. Patricia at Detector said her company would refuse to do worker tracking, stating that 'we want to preserve ethics very much'. Even Gareth at Rely, who was quite dismissive of some of the ethical issues that I put to him, concurred. He said:

> So if a brand wants to know how many people talked about our brand, and we give them the answer of 500, so what? But if someone wanted to do surveillance, what I would say, 'Well, that's not really what this has been designed for, you go off and do that yourself, that's not something I will support.'

Thus most respondents showed that there was an ethical line they would not cross, although not all did this. We have seen, for example, that one of the companies in which I carried out an interview, Discern, carried out worker tracking for a multinational client. The location of that ethical line, then, differs because of people's different conceptions of what is ethical and fair. They make different decisions about what is ethically acceptable, and these play a role in shaping what data mining gets done and for what purposes.

Mark Andrejevic claims that the abstraction of emotions from individuals through sentiment analysis, one particular form of data mining, plays a role in controlling affect, which he describes, quoting Massumi, as 'an intrinsic variable of the late capitalist system, as infrastructural as a factory' (Massumi 2002, p. 45, quoted in Andrejevic 2011, p. 609). Andrejevic argues that affect, 'a circulating, undifferentiated kind of emotion' (2011, p. 608), is an exploitable resource within affective economies, and its exploitation results in forms of control. So, for him, abstracting emotions from individuals and aggregating them for the purposes of prediction and control is problematic, whereas for the practitioners to whom I spoke, this distance from the individual is a kind of ethical safeguard. For many, keeping data mining general and not attending to the individual was a more ethical practice, as seen in Bernhardt's comments above about his company's preference for capturing 'the general feeling' rather than focusing on individuals. This view was echoed by an academic social media data miner who I spoke to early in my research, who said:

> I don't monitor people, in the sense that nothing that I do is intended to affect the people that I'm getting the texts from. So the end result of my analysis is completely neutral to them. So I won't try to sell them anything, or try to get them to modify their behaviour in any way.

Transparency as Ethics

The above quote points to another ethical issue: users' awareness of social media data mining practices. In my focus group research with social media users, they did not believe, as this academic does, that the most ethical approach to data mining is for users not to know that it is happening. In my interviews with commercial insights workers, Robert from digital marketing agency BrandHook felt that user awareness, or lack of it, was the central ethical issue in relation to social media data mining, raising the question of who is responsible for facilitating more awareness and greater transparency. Opinions among respondents differed. Most participants believed that social media platforms have the responsibility to be transparent about the data mining that they permit. Daniel, from the reputation management company Checker, who described himself as having a background in 'integrity-led' corporate communications, was particularly vocal in his criticism of social media platforms. These platforms benefit from the ambiguity regarding what is public in social media and from the absence of appropriate legislation, he claimed, and they use the absence of legal liability as an excuse for failing to exercise due diligence. It is here that critical attention needs to be focused, he argued, rather than on data miners themselves. Isabel at the digital marketing agency Octopus pointed out that platforms have known arrangements with social media insights companies (such as the historical relationship between Twitter and Radian6, in which the latter had access to the former's firehose of data), so they cannot claim to be ignorant of the monitoring that takes place. This heightens their responsibility to act more transparently, she suggested.

Some respondents said that social insights companies also had a responsibility to make data mining transparent. Gareth at Rely felt that all parties shared this responsibility: social media platforms from which data are extracted; companies like his which gather and monitor data; and the companies that purchase the data and services of the intermediary insights sector. Gareth was proud of his own company's approach to explaining transparently how they calculate reputation, stating:

> We have a part of the website which explains in detail how we score. Every single one of the profiles has an activity statement and you can see every single tweet and how many points we've awarded. […] We're very open when we deal with clients, we share data. […] Even if people don't like what we do or don't agree with how we score, we can stand up and say, 'Well, at least you know how we score.' […] And I've actually won deals, where I've won against a competitor who wasn't being as transparent, and I was told, 'The reason we gave it to you was because of your transparency.' So I did that because of who I am, not because I want to win a deal, and it's pleasing to know that by being transparent you can be the nice guy and actually win. And so for me, one, I'm transparent, but I keep doing that because I know that commercially it benefits us.

As with the decision to access only public data, apparently moral positions like Gareth's clearly have an economic advantage. So what might seem ethical is also driven by business interest. Gareth went on to say that he did not want regulation to require all players to be transparent, because his company was currently benefiting from being more transparent than others. But he later backtracked, saying he would like to see other companies acting more transparently. Isla at the media monitoring company BlueSky also did this, saying first that it was individual social media users' responsibility to know that data mining happens (by reading terms of service carefully, for example), and later that companies need to do more, citing Instagram's changed terms and conditions as an example of bad transparency management. These contradictory things that respondents said about transparency point to the ethical ambiguity of the work of social media data mining.

As social media data mining becomes ordinary and commonplace, new relationships around data emerge, and these, in turn, require a new ethical framework which is appropriate to these emerging relationships. In the differences, ambivalences and contradictory remarks about ethical issues identified here, we see uncertainty with regard to what that framework might look like. These workers sometimes appear unsure of where to get their ethical ideas from, or on what discursive repertoires to draw in order to express their ethics. As a result, their comments sometimes reflect company policy—their job is to fulfil the desire for numbers, and to some extent this obliges them to toe the company ethical line. Yet at the same time, they display personal ambivalence towards such policies, depending on their own comfort or discomfort with digital data tracking.

Who Benefits from Social Media Data Mining?

Amanda displayed similar ambivalence to Gareth and Isla—first sure, and then not so sure—when it came to the issue of who benefits from social media data mining. At first, Amanda pointed to the ways in which social media users might benefit. Initially, she noted that her company gathers data for a range of different organisations, some of them acting in the public interest. 'Charities use social media monitoring to understand how people feel about world events and how they can encourage them to contribute to positive change by donating money or volunteering time', she said. Other respondents also pointed to similar benefits. For Bernhardt at the media monitoring company Claimr, the majority of his company's clients are media organisations, interested not in growing brand equity or changing consumption behaviour, but rather in understanding people's concerns.

The sentiment analysis company SentiCheck works extensively with a world leading cooperative which promotes ethical business practices. My interviewee there, Graciela, felt optimistic about the predictive capacities of sentiment analysis which concern Andrejevic, because they may contribute to the development of treatments relating to emotional ill health such as depression, she suggested.

Lawrence at the insights company 24-7Social pointed out that even mining data on behalf of powerful companies might have social benefits. He told this story about the work his company did for a multinational oil company after it had experienced a major spillage:

L That was quite interesting, because obviously there was a degree of, basically a protest against them. To be honest they were too busy to try and do anything about the protest, but there was also definitely a clean-up aspect to it, in that someone would tweet, 'Just found a dead seal here?' and they could get their dead seal collection unit to go and pick up the dead seal and scratch some of the oil off.

HK So it was helping them to clean up?

L Yes, so even though people would say, do you want (the oil company), the most evil company on earth, monitoring what you are saying about the oil spill, the appalling natural disaster they have just created, at the same time monitoring enabled a positive outcome, at least one aspect of it did.

Some respondents claimed that because of the opportunities that social media data mining opens up for ordinary voices to be heard by brands, consumers now have more input into how brands operate. Popular examples of such benefits include stories about the companies Dell and Motrin (Hunt 2009). In response to negative blog posts about poor quality customer service and fearful of losing market share, Dell set up Direct2Dell to encourage customers to share their frustrations directly with the company. Initially, they received many negative comments, but their responses to these comments were seen to rebuild trust with customers over time. In another incident, painkiller Motrin released an advertisement targeting mothers carrying their babies in carriers, which was not well received because of its flippant tone. Negative commentary spread rapidly on Twitter, after which Motrin removed the advertisement and apologised. 'People have more power to impact on how companies behave', said Amanda from Witness in relation to these kinds of examples. Bernhardt from Claimr said that consumers refuse to behave on social media platforms as companies would like them to, which he saw as a form of consumer power. 'I don't see that trying to catch the sentiment from these kinds of data is something that empowers companies more than normal persons', he proposed, because consumers have the power to say 'we choose to say something opposite to what they want us to say'.

While such views may seem to reflect idealised notions of the empowering potential of participatory cultures which we may want to problematise (and indeed some respondents did just that), they also point to a belief in the possibility of social media users' agency in the production of knowledge about them. Academic research into social media usage also points to the agency of social media users, who make conscious decisions about which views to express or perform, and who are thus not simply passive victims of data mining (such as Marwick and boyd 2010). But some respondents expressed concerns about whether consumers do actually benefit from the mining of their social media data. Amanda at Witness again, doubting her own assertions about the benefits of commercial data mining, said:

> Say for example a company might want to know why don't parents invest in our brand, and they might greenwash, saying they're environmentally friendly without making any changes, as opposed to actually making change. If the outcome of the work we're doing is that companies understand better how to message things to people, I'd probably be very uncomfortable. I would be very uncomfortable with it.

Gail at Paste put it much more starkly:

> The ability to be able to track behaviour only really helps the person who wants to track that behaviour and not the person who is being tracked. The kind of idea that they might be given a better experience, or better products, I just think is a load of rubbish.

Regulation as Ethical Solution?

Discussions with interviewees about whether and how the sector might be regulated highlighted the tension between the moral and the economic, and between interviewees' identities both as social insights professionals and as social media users. Respondents had mixed views about whether regulation was needed. Some, like Gareth at Rely, felt that it was not necessary because common sense could be trusted to prevail. Alexander, at Fusage, felt that 'cowboys', as he called them, or people engaged in bad practice, would disappear, as, with time, clients would understand social insights processes better and so be able to ask better questions of the companies they engaged to do this work. Nonetheless, said Alexander, a body like the Internet/Interactive Advertising Bureau (IAB) might be needed to ensure good practice. Isabel at Octopus summed up what she saw as her own contradictory responses to this issue, identifying that regulation might simultaneously help users and hurt insights companies:

> If we were regulated more on data protection or from what we were monitoring, it would hinder the work that I'm doing or make my job a little bit more difficult, I guess. But with my personal head on, I think when people should be thinking about what they're tweeting, then yeah it should be governed. I kind of have two different heads here.

Robert, at digital marketing agency BrandHook, brought up the topic of regulation, rather than waiting for me to ask questions about it as other interviewees did. He spoke at length about Facebook and Twitter's data ownership, the lack of appropriate regulation relating to this, and the need for a regulatory body to be established. He said:

> I think the problem currently with social media is that it's not very regulated. So with brands and agencies and professionals, we tend not to really question it too much, because we're waiting for somebody to say you can't do that and we go okay. […] The ethics of data is really difficult and it's a debate that needs to go on, but I think there needs to be some regulation or some sort of agreement, because people are always going to abuse it until they feel there's some sort of threat, that if you're found to be doing this we're going to take you out.

Efforts to explore how the sector might be regulated have started to emerge. Some of these originate in the sector itself, building on guidance and codes of conduct developed, for example, by the IAB, the Advertising Standards Agency (ASA), or the Market Research Society (MRS). In the US, the Big Boulder Initiative, a not-for-profit trade association established in 2011 with the aim of ensuring the long-term viability of the social data industry, has produced a code of ethics which attempts to 'define a set of ethical standards for the treatment of social data' (http://blog.bbi.org/2014/11/14/draft-code-of-ethics-stan-dards-for-social-data/), which relate to: accountability, privacy, transparency, education and accessibility. WOMMA, the Word of Mouth Marketing Association in the US and the UK, which describes itself as 'the leading voice for ethical and effective word of mouth and social media marketing' (WOMMA 2012) offers, among other things, ethics codes and ethical resources, such as social media disclosure and privacy guides, guides relating to honesty about ROI, and ethical assessment tools. The IAB, which describes itself as 'committed to the continued growth of the interactive advertising ecosystem in tandem with ethical and consumer-friendly advertising practices', has a self-regulatory programme for online behavioural advertising, which involves displaying an icon on webpages where data is collected for these purposes (http://www.iab.net/media/file/OBA_OneSheet_Final.pdf). But while some discussion about regulation and ethics is emerging, there is still a long way to go. At Social Media Week in London in 2014, I attended a presentation about social insights work, surveillance and ethics, given to a near-empty room. For the following talk by BuzzFeed, the room was full—perhaps of people hoping to learn how to market their start-ups to venture capital investors. A year earlier, I attended a Westminster Forum (http://www.westminsterforumprojects.co.uk/) event about policy priorities for social media, described as addressing data mining among other things, at which none of the speakers discussed regulating social media monitoring. When I asked a question about how it might be done, the expert panel, which included representatives from the ASA and social insights companies, appeared surprised at the suggestion that the sector—not broken, in their view—might need to be 'fixed' through regulation. In light of such inaction, data activists and other groups are lobbying for legislation and policy in relation to digital data tracking and data rights, something I discuss briefly in Chapter 8.

CONCLUSION: CONCERNS AND ETHICS IN COMMERCIAL SOCIAL MEDIA DATA MINING PRACTICE

Intermediary commercial social insights companies have distinct origins and offer distinct services: some offer full digital marketing and some limit their services to data mining and reporting, leaving it to their clients to decide what to do with generated data. Within this sector, data mining is not all the same, and we need to attend to these differences in order to develop understanding of what it is and how it gets done. But all companies grapple with the quality and accuracy of social data and the relationship between this (lack of) quality and clients' desires for numbers. When it comes to the economic value of social media marketing and data mining, clients demand precise figures, but when it comes to numbers relating to mined social data, there is less demand for precision. In the last chapter, I argued that this desire for numbers, produced by datafication, big data rhetoric and engagement in data mining made it difficult to reflect critically with public sector partners about some of the problems that working with data mining entails. In this chapter, I noted that although many interviewees recognise data to be volatile, inaccurate and messy, their clients' desire for numbers, accurate or not, also made it difficult for insights workers to communicate the limitations of the numbers they produce to their clients. So although interviewees show awareness of the methodological and epistemological issues that accompany data mining, they felt restricted in their ability to talk to clients about these issues. Clients' desire for numbers, any numbers, suppresses debate about their volatility and fragility. What is lost through this desire for and dependence on quantities, therefore, is understanding of them and the things they purport to represent. As noted in the previous chapter, quantification is a way of managing the world, not understanding it, and so when numbers dominate, the understandings that qualitative sensibilities make possible are absent (Baym 2013).

In this chapter, I highlighted the ways in which insights work simultaneously involves ethical and financial considerations. The decision to access only public data, however ambiguous this concept might be, and the commitment to the idea that being transparent about data mining is a good thing, are both moral and economic in character. Decisions about what kind of data to access that are led by economic concerns can also be morally advantageous, and the decision to be transparent about how data is mined that is based on ethical considerations can also be financially beneficial. The opposite is also true. Interviewees showed that they have individual, ethical barometers; they draw lines around what they will and will not do, but these lines are not in the same place for all workers. Some interviewees drew attention to the ethical dilemma raised by their contradictory identities as insights workers and social media users when they acknowledged that regulation that protects them as users might hinder them as workers. Some respondents believed that consumers benefit from social media data mining, whereas others recognised this as rhetoric which serves to obscure who really benefits. In practice, commercial ethics are contingent, variable, individualised and unstable. We might characterise the ethics of commercial social media data mining as being in a state of interpretative flexibility, a term used within Science and Technology Studies (STS) to characterise socio-technical assemblages for which a range of meanings exist, whose definition and use are still under negotiation (Law 1987; Wyatt 1998). As social media data mining becomes ordinary, new data relations emerge, and there is uncertainty about what kinds of ethics should be applied to these relations. As a consequence of this uncertainty, the individual decisions of data workers shape how data mining gets done and how it gets stabilised. Interviewees' histories inform their decisions, from their educational backgrounds to their experiences as workers, volunteers and human subjects. But so do company policy and the fact that it is their job to contribute towards fulfilling their clients desire for numbers. So, despite some acknowledgement that 'public' is not a straightforward category in social media, most respondents nonetheless adhere to the mantra that if social media data is public it is therefore 'fair game' to be mined and analysed. And, despite commitment to the ideas that greater transparency about social media data mining is needed and that regulation is one way to ensure it, interviewees were not actively engaged in attempting to bring about better, more transparent practices.

So although it is encouraging to see ethical reflection among these workers and to note that some practices are off-limits to them, critical methodological knowledge and ethical thoughtfulness are often overshadowed by commercial imperatives and the desire for numbers. Interviewees have thought about some of the ethical questions that these methods raise, more so than some of the participants discussed in the previous chapter—not surprisingly, because they are professional data workers. Understanding ethical decision-making and line-drawing as acts of worker agency, we might argue that workers can and do act with agency in relation to their data mining work. They do not simply submit to or reproduce its 'harsh logic' (Feenberg 2002, p. v), even if its harsh logic retains dominance. At the same time, we might see the contradictions within what interviewees say, as they backtrack and change their minds, as 'ethical openings' (Gibson-Graham 2006, p. 135), cracks in the armour of corporate discourse, in which thinking through the ethics of data mining might take place, albeit within the constraining conditions of desiring numbers.

Some issues which have surfaced in this chapter are discussed in more detail in later chapters. One is user understanding of and attitudes towards data mining, which I return to in Chapter 7. Another is that interviewees and their companies often do not know what happens as a result of the data mining that they do—in some cases this is because companies provide data reporting services, handing data over to clients who then decide what to do with it. Occasionally, interviewees could give examples of actions taken. Lawrence from 24-7Social, for example, reported that his company identified a planned demonstration by the Occupy protest movement outside a client financial organisation, which enabled the client organisation to take steps to secure its property and send workers home for their safety—an example which raises all sorts of ethical issues beyond those discussed here. Online, company websites emphasise the positive outcomes of doing social media data mining, through case studies and statistics relating to increased reach, yet sometimes workers in these companies would like to know more about the consequences of their data mining work. Alexander, from the insights company Fusage, said:

> It's probably one of the frustrations of being a researcher that I've always found in my life, is you get so far and then you almost wave bye to it. Because then the agencies take over the implementation and the brand marketing teams take over and use that insight to do whatever they do. So we don't, if I'm honest, always see the fruits of our labour in terms of strategy or in terms of the implementation thereof.

My interviewees are not alone in not knowing what happens to mined social media data; there is little discussion of the actual, specific consequences of social media data mining, especially when it is undertaken by ordinary organisations. Knowing more about consequences might help to address the question of what should concern us about it. The next chapter focuses on what happens to mined data, and whether it should worry us.

CHAPTER 6

What Happens to Mined Social Media Data?

INTRODUCTION

In this chapter, I focus on the consequences of organisational social media data mining. The criticisms of social media data mining which I discussed in Chapter 3—that it invades privacy, expands surveillance to personal and intimate realms, results in new forms of discrimination and all kinds of exclusions, some new and some old—depend, in part, on what happens as a result of data mining and its concrete, material consequences. Noting McStay's assertion that accounts of data mining which confuse actual and potential practices 'serve only to foster Big Brother-type phantoms' (2011, p. 85), and in order to address this book's central question of what should concern us about social media data mining, we need to attend to its outcomes. In making this argument, I return again to Couldry and Powell's assertion that we need to ground the study of data practices in real-world, everyday contexts, or in 'what actual social actors, and groups of actors, are doing under (datafied) conditions in a variety of places and settings' (2014, p. 2).

There are plenty of stories in the public domain about the consequences of modern-day data gathering that would lead us to conclude that this is a phenomenon about which we should be concerned. These stories tend to focus on the highly visible practices of large-scale organisations, the major social media platforms and government and security agencies, as noted in the introduction to the book. What is missing from this picture, as I have suggested, are more ordinary forms of social media data mining. What

H. Kennedy, *Post, Mine, Repeat*,
DOI 10.1057/978-1-137-35398-6_6

129

happens to social data gathered by a museum, a council or a university, and what are the consequences of the data mining and analysis undertaken by such organisations? We need to know the answers to these questions in order to be able to assess the issues at the heart of this book. This chapter attempts to address those questions, drawing on interviews with organisations which have used the services of social media insights companies or undertaken their own social media data mining.

Getting organisations or their employees to agree to be interviewed about their data mining is not easy, because public concern about things like privacy and surveillance can mean that organisations are guarded about speaking openly about the analytics activities that they undertake. Partly because of this guardedness, it can be hard to identify the right people to approach about these matters—organisations often do not share details of named individuals responsible for mining data. As a result, it can be difficult to secure interviewees. I managed to persuade people with data mining remits in ten organisations to agree to participate in interviews, in which I asked about: their background and role; the social media data mining and analytics undertaken within the organisation; the purposes for which such work is undertaken, the uses to which data are put and the concrete outcomes of data mining; and issues in and barriers to doing data mining. I also asked interviewees if they would share documentation with me which provided evidence of action taken. Half of the interviewees did this.

Drawing on these interviews, I make four main points in this chapter. First, the concrete changes that result from data mining are rather limited. Often, social media data mining confirms the value of organisational participation in social media communication and engagement and, as such, confirms the value of social media data mining itself. In other words, the object of data mining is sometimes data mining, and measurement is sometimes undertaken in order to confirm the value of measurement. Changes that result from data mining include things like buy-in to data mining by senior colleagues and subsequent minor organisational shifts to accommodate data mining practices. Such mundane and intra-organisational change seems modest and unexciting, and does not appear, at first sight, to be concerning. However, the second section of the chapter argues that such changes should concern us, because modifications to working arrangements made as a result of these internal, organisational changes have serious consequences for the working life of staff within the organisations. More social media engagement and monitoring means that staff are required to work at times considered to be optimum for social media engagement and monitoring, including early mornings, evenings and weekends. They experience 'function creep', as Gregg (2011) calls it, in that the boundaries around workers' roles expand to include cycles of social media engagement, analytics and improved engagement. What's more, employers track and measure worker performance through the same data mining procedures that workers themselves are increasingly required to implement.

In the third section of the chapter, I discuss the attitudes to statistics, data and other quantities that informed the data mining experiences of respondents. I show how the data evangelism of people within organisations undertaking social media insights work can lead to an overstatement of the extent of change that results. Interviewees either revealed their own evangelism or reported similar faith in metrics among their less insights-experienced colleagues. This was frequently expressed as what Grosser (2014) describes as a 'desire for more'—more followers, more retweets, more shares, higher numbers. Here again we see how datafication, the rhetoric of big data and engagement in data mining combine with a more historical trust in numbers' proclaimed objectivity to produce a prevalent desire for numbers. As noted in the previous chapter, this desire serves to overshadow consideration of ethical issues relating to data mining, such as privacy, security and transparency. On the whole, interviewees gave conventional answers to questions about ethics, subscribing to the view that social media data are 'fair game' for mining, noting that their companies adhere to data protection measures and pointing to their terms and conditions and privacy statements as evidence of their 'good practice' with regard to transparency.

Through discussion of these points, this chapter highlights the complex conditions within which ordinary organisations attempt to adapt to social media data mining as it becomes ordinary. On the one hand, these complex conditions mean that there is little concrete change outside of the organisations themselves, which suggests that there is not much to be concerned about. But on the other, organisational changes to working practices and the predominance of particular attitudes to statistical quantities are concerning in terms of quality of working life and broader cultural changes that occur when a desire for quantities becomes dominant. I return to these points in the conclusion to this chapter. In the next section, I describe my interviewees, their organisations and their uses of social media data mining.

INTERVIEWEES, ORGANISATIONS, AND THEIR USES
OF SOCIAL MEDIA DATA MINING

In the first half of 2014, post-doctoral researcher Stylianos Moshonas and I carried out interviews with employees from ten different organisations, mostly UK-based; I use pseudonyms to refer to them in this chapter. I classify the organisations into five types, in each of which we carried out two interviews: universities, media organisations, local councils, museums (not the same local councils and museums with which we undertook the action research discussed in Chapter 4) and non-profits. We interviewed Jane, Head of Digital Communications in University 1 and William, Social and Digital Media Officer in University 2. Jane leads a team which manages and develops strategy for the university's website and social media channels, which target a broad range of audiences, from prospective students and staff to international research communities. Similarly, William is responsible for his university's presence on social media platforms, including Sina Weibo in China, and Chinese video hosting service Youku. With this global social media presence, according to William, his employer is acknowledged as a leader among UK universities in its social media engagement. We also carried out interviews in two media organisations, one UK-based and one global. In the former, Media Organisation 1, we interviewed Ash, Head of Data Planning and Analytics, whose role is to manage the extraction of value from data across all parts of the business (sales, commissioning, scheduling, marketing, CRM (customer relationship management), content management, and rights management). In the latter, Media Organisation 2, we interviewed David, General Manager of the company's international arm, responsible for all digital products outside the USA.

In the local councils, we interviewed Anthony in Council 1, a Senior Digital Marketing Manager in charge of a small team, and Ruth in Council 2, Digital Marketing Officer responsible for her employer's advertising campaigns on digital and social media channels. We interviewed Natalie, a data analyst at a major UK museum, who had total responsibility for analysing and evaluating traffic through digital channels, including the museum's website, social media accounts, and mobile and interactive applications. At the time of the interview, she stated that she was the only full-time data analyst employed by a UK museum. The other interview we carried out within the museum sector was with two researchers in The Netherlands, Jan and Michael, who had developed a social media data mining tool for museums to compare themselves to one another and to access social media data from a range of platforms. Finally, we carried out two interviews in non-profit organisations. This latter category included a member-owned family of businesses, Non-profit 1, and a professional accountancy training association, Non-profit 2. In the former, we interviewed Cathryn, a member of the four-person social media team, a sub-section of the marketing team, which sits at the heart of this structurally complex organisation. In the latter, we interviewed Suzanne, the organisation's Social Media Manager, based at its headquarters in London; the organisation has over 35 branches around the world. Table 6.1 provides a summary of interviewees and their organisations.

This small sample of interviewees represents a modest cross-section of organisations attempting to adapt to a changing social media and big data landscape, and exploring different approaches to data mining in order to identify strategies relevant to their organisational objectives. While not all equally ordinary—the media organisations and the UK museum are more high profile than the local councils and regional universities, and as a result, are more advanced in their data mining practices—together they form much of the infrastructure of everyday life, covering the domains of education, media, culture, public services and leisure.

The organisations use a diverse range of social media data mining tools and processes. University 1 uses Hootsuite, a dashboard for managing multiple social media channels, to produce monthly reports of social media activity and campaign-based reports. It also uses Facebook Updates and Google Analytics, this latter for web analytics and to understand the traffic between the web and social media. This university recognises that Instagram is an important platform for engaging prospective students, but as the platform has limited analytics functions, not much mining is possible therein. University 2 also primarily uses Hootsuite, but subscribes to a corporate package, making their use of it more extensive and sophisticated than University 1. Hootsuite is used to monitor relevant stakeholders, including students, via hashtags and search phrases relating to the university. William's team also uses GnATTERbox, a tool that alerts his team to mentions of his organisation by named politicians, journalists and business leaders. Finally, they carry out manual monitoring on the Student Room, an online discussion space about student issues.

Table 6.1   Interviewees, their organisations, and social media insights tools used

Interviewee	Organisation	Organisation details	Insights tools used
Jane, Head of Digital Communications	University 1	Based in a northern city	Hootsuite, Facebook Updates, Google Analytics
William, Social and Digital Media Officer	University 2	Based in a northern city, UK leader in social media engagement	Hootsuite Pro, GnaTTERboox, Student Room
Ash, Head of Data Planning and Analysis	Media Organisation 1	UK, London-based	Primarily bespoke dashboard, also Second Sync, previously Sysomos, Radian6
David, General Manager	Media Organisation 2	Global; David is responsible for the company's international, non-US operations	Adobe Omniture, Visual Revenue, Outbrain, ChartBeat
Anthony, Senior Digital Marketing Manager	Council 1	Northern county council	Meltwater Buzz
Ruth, Digital Marketing Officer	Council 2	Junior staff member in northern city council	Meltwater Buzz
Natalie, Data Analyst	Museum 1	Also a researcher; London-based	Google Analytics, Hootsuite, RowFeeder, Crazy Egg, Facebook Insights, Twitter Analytics, YouTube Analytics, Flurry, bespoke dashboard
Jan and Michael, researchers	Museum 2	Developed social media mining tool for museum sector; Netherlands-based	Created bespoke dashboard for museums sector
Cathryn, Social Media Team member	Non-profit 1	UK-based, member-owned family of businesses	Meltwater Buzz
Suzanne, Social Media Manager	Non-profit 2	Global professional accountancy training association; headquarters in London	Bespoke analytics dashboard, Conversocial, previously employed in social insights company

Both of the councils and one of the non-profit organisations in which we carried out interviews use Meltwater Buzz (we were put in touch with these organisations by a contact at Meltwater). They use it to find out what people are saying about the organisations on social media, to help them schedule the distribution of social media messages from multiple accounts and to manage and track keywords from multiple branches. Non-profit 2, the professional accountancy training association, previously employed social media insights companies who used third party social insights tools to monitor buzz on relevant topics but, at the time of the interview, they were undertaking their own analytics, having developed a purpose-built dashboard with the help of the insights company. This organisation employs an in-house insights specialist to carry out its analytics. It also uses Conversocial, a web application embedded in Facebook that allows CRM agents to reply to users in real-time and track responses.

As can be seen, a combination of ready-made and purpose-built tools is used by these organisations for data mining purposes. This was the case at the UK-based museum, where our interviewee Natalie described data mining as somewhere in between a settled process based on established and known tools, and an experimental approach which involved trying out new tools to meet specific needs. Tools used include Google Analytics, Hootsuite, Row Feeder (a Twitter tool used to analyse hashtags for specific exhibitions), Crazy Egg (a heat map tool that shows when people click on the website), Facebook Insights, Twitter Analytics, YouTube Analytics, and Flurry for mobile applications. This palette of tools was used because Natalie could not identify one which provided all of the insights that the museum required. The tool produced for museums by Dutch researchers Jan and Michael represents an attempt to fill this gap, and Natalie also attempted to do this for her organisation by developing a bespoke dashboard which compiles data from web, social media and mobile applications.

The two media organisations also use both custom-built and off-the-shelf tools. Media Organisation 1 had previously experimented with tools like Sysomos and Radian6, but found these did not meet their needs when it came to analysing social data about multiple brands or hierarchies of brands (such as channels, programmes, genres, talent). At the time of the interview, they were using Second Sync, a Twitter analytics tool used by media production companies and broadcasters, which allows them to focus on specific shows and content, rather than general conversation, and to 'overlay the conversation with the linear transmission', as Ash put it. This organisation is rare in that most of the data mining it does is made transparent to audiences: it invites viewers to register and voluntarily share their personal data, which then form the basis of analysis. The other media organisation uses the Adobe Omniture platform for real-time analysis of the organisations' website and social media channels. It also uses Visual Revenue to test headlines, assessing the relative success of different headlines and adapting headline-writing practice according to outcomes. Outbrain provides them with data about the sharing and recirculation of online content, and Chartbeat, a dashboard-based service, provides them with real-time data about likes, shares and so on.

According to our interviewees, these various tools are used in order to:

- complement traditional means of gathering feedback (such as surveys and focus groups) and find out what people think;
- listen for and audit mentions of the organisations, specific campaigns and aspects of their work, and find out what is trending, which would not be possible through traditional means;
- use this information to help develop digital/social media engagement strategy, to inform decision-making, to help the organisation reach a range of objectives;
- build a picture of who users are and understand how social media can be used to reach new audiences;
- assess performance, identifying where improvements can be made in communications approaches, justifying activities with quantitative data;
- build predictive models of audience preferences to personalise experience, target advertisements, amplify stories;
- improve user experience, ROI and efficiency in times of cuts;
- do CRM;
- identify optimum times of the day for social media engagement;
- manage crises and track related sentiment;
- highlight the value of social analytics and insights within the organisation.

To what extent are organisations successful in meeting these objectives? What happens to the social data that are gathered? Does data mining result in concrete change? The next section addresses these questions.

THE CONSEQUENCES OF SOCIAL MEDIA
DATA MINING

Concrete Action and Organisational Complexity

I asked interviewees to give examples of concrete action and change that resulted from the social media data mining undertaken within their organisations. They had surprisingly few examples to share, and those they had related largely to internal, organisational change. Perhaps because media organisations expect to engage with their audiences through social and other digital media, the most advanced examples of concrete change came from them. The UK media company, Media Organisation 1, targets content on its website based on the understanding of its audiences that it derives from data mining and the predictive models that it has subsequently developed. These models, our interviewee Ash claimed, have been validated by an external company as 90% accurate, which has enabled the organisation to sell advertising at a premium rate and thus generate income, which is invested back into content production activities. The global company, Media Organisation 2, has changed a number of practices as a result of data mining. Analysis suggested that their audiences were more interested in 'what is coming next' than 'breaking news', so the organisation changed its editorial strategy to include more commentary and reflection in news stories than was previously the case. In another example, the organisation used social media data mining to assess how different headlines perform, in order to ensure that headlines attract maximum audiences, and changed their headline-writing practice accordingly. The findings of data mining have led to changes in what is expected of journalists, who now have to meet targets such as increases in audience numbers, pages surfed, pages viewed per user and per visit. To achieve these targets, journalists must come up with amplification strategies that work, for example identifying which Twitter hashtag makes a story most effective. Thus journalists are expected to codify mined data themselves.

Other organisations identified damage limitation as a concrete action resulting from data mining. Anthony in Council 1 provided some examples. He told us that late one Friday afternoon, his social media analytics team noticed residents in a particular area tweeting that they were getting spots of oil on their cars. His team investigated, and found that there had been an incident at a nearby oil refinery. Because the council identified this problem in real-time, it was able to quickly put procedures in place to deal with the incident: it let its contact centre know to expect an increased volume of calls, spoke to the refinery to get some facts from them, got the message out to the public to explain what had happened and reassure people that the oil was not toxic, and offered residents a free visit to a local car wash, courtesy of the refinery. Anthony said:

> It was the social media spot that actually triggered it. It would have hap-pened, you know, all that would have happened normally but it would have happened a lot later. It would have happened hours later, probably. It happened at three or four o'clock on a Friday afternoon; if we'd have started receiving those calls at half-five, six o'clock on a Friday afternoon, the chances of us being able to get so many people mobilised so quickly would have been diminished.

Non-profit 1, the family of businesses, also provided examples. One action they took was to change their chicken stocking density (that is, how closely chickens are kept together in a coop) as a result of concern expressed by customers on social media about chicken welfare. In another example, the organisation was tagged in a photo on Instagram of a sign in a local shop stating that 'Ass Cream Biscuits' were on sale, with 'Ass' here an unfortunate abbreviation for assorted. Our interviewee Cathryn acted quickly to com-municate with the signage department in order to have the sign removed.

In these examples, concrete actions relate to organisational change— different headlines, different expectations of journalists, quicker responses to potentially damaging situations. Other organisational changes also took place. For example, as a result of finding out through social media data mining that social media communication is effective, some organisa-tions, such as the local councils, decided to implement 'channel shift'— that is, shifting from print-based communication to digital and online approaches. Others, such as the non-profits, changed the scheduling of their own social media activity on the basis of their analysis of optimum times to engage with their publics via social media platforms. Cathryn in Non-profit 1, said:

> Now we're posting before work, so like in commuter downtime, so about 7.30 am is a really great time for us to post. And then in the evening about 7 pm we're posting, and then on weekends at about 10 am. And that's kind of a change in our behaviour, from looking at the insights. [...] Like we're meant to work 9–5 as a team, but social media is not so 9–5. So we've changed the hours that we're working as well, to adapt to that.

Anthony in Council 1 reiterated this point, stating that 'even on Good Friday, you know, there's someone at the other end who will be going through this and who will be actually listening, acting and taking away what people say. Nobody speaks into a vacuum when they speak to us.'

Other organisations have made the decision to dedicate more resources to social media data mining as a result of the insights that data mining has produced. Non-profit 2 decided to engage actively in social media communications and recruit new staff, including an analytics expert and it now encourages staff in its 35 offices around the world to get involved in social media, as their analysis showed them that there was appetite within their markets and audiences for them to do so. A set of Key Performance Indicators (which include measures such as likes and reach) has been developed to assess social media performance, and the activities of social media engagement managers in regional offices are evaluated against these indicators in monthly meetings which examine performance and potential improvements.

Non-profit 2 is a good example of an organisation in which concrete changes resulting from social media data mining relate largely to internal company operations and to social media themselves. Analytics are undertaken with the aim of persuading managers of the value of investing in social media marketing and communication. Thus social media data mining is deployed to prove the importance of social media and of data mining. One way that this is done is through the rhetoric of cost-effectiveness. Cathryn in Non-profit 1 said that data mining can prove that marketing on social media has a much greater reach than other, print-based forms. Having statistics that demonstrate the value of £1000 spent on social media compared to other media, she said, makes it possible to increase social media budgets. Another interviewee, Anthony in Council 1, claimed that he would have to employ 3.5 full-time staff to do the equivalent of automated analytics work, and he calculated that he was saving his employer a five-figure sum by using data mining methods.

It was surprising how few concrete examples of action and change were provided by interviewees. This happens, in part, because the kinds of actions to which mined data are amenable are often minor organisational adjustments. Even Non-profit 1, which, at the time of the interview, had recently experienced a PR crisis, used data mining about this crisis to persuade others in the organisation that data mining is worthy of investment. Our interviewee Cathryn said of their crisis-related data mining:

They say, 'Oh, actually, that's really useful, to see what customers have said', because I did a graph (with) the volumes. [...] You hear, 'Oh, customers are saying', but seeing it all together, I think it gave a bit of context to what everyone was hearing lately. And I think that engaged a few people. They thought, 'Oh, yeah, social media can actually tell us', because it's so instant, it can tell us about last week and the week before, in a way that they've not really seen that anywhere else. So, yeah, [...] I think getting that current data and customer feeling, that probably engaged a few different people.

But despite the modest scale of organisational adjustments, interviewees did not identify changes as either few in number or internally focused. On the contrary, most spoke passionately about what was possible with data mining. William in University 2 was an exception, as he acknowledged that the results of his organisation's data mining were limited and largely informative, circulated in order to 'keep core staff in the loop' and to reassure them and others 'that everything is ticking along'. Other interviewees acknowledged that, although concrete action had taken place, more action was possible, and most recognised that organisational complexity was a barrier to implementing change. Some organisations were amenable to incorporating social data mining into their practices and adapting as a result of insights, but others were structurally more inhibiting of such change. Organisations are not equal in this sense, an important fact in considering 'the actual processes of data-gathering, data-processing and organisational adjustment' (Couldry and Powell 2014) associated with data mining. In Non-profit 1, Cathryn told us that her organisation's structure made it hard to get insights to the right people. She said: 'it's hard for us to know who's the right person when there's, like, thousands of people. So I think the way the organisation is structured is a limitation.' This is exacerbated by the absence of established procedures for integrating social insights into organisational workflows, which exist for things like telephone and email communication. Not all data are relevant to all people in all offices, said Suzanne at Non-profit 2, making it difficult to share insights effectively, and interviewees in the UK museum, Council 1 and both universities also saw structural complexity as a barrier to turning insights into action. Thus the ability to enact change as a result of data mining is constrained by complex organisational arrangements. As a result, most changes that resulted from social media data mining in the organisations where I carried out interviews were internal to organisations. I discuss the implications of this phenomenon in the next section.

Organisational Change and the Quality of Working Life

Not much concrete change took place within interviewees' organisations as a result of social media data mining. Where concrete actions were identified, some were less concrete than others, and many related to making changes to how organisations do data mining, such as appointing a dedicated data mining expert, or how they use social media, such as changing timings of engagement and making more widespread use of social media within their communication strategies. There seemed to be little change to the external services that the organisations offer. All of this would seem to suggest that there is little substantive consequence of these ordinary organisations undertaking social media data mining and that, therefore, there is little to concern us here.

However, these apparently mundane changes do raise important concerns, and not just in relation to the issue of exclusion—that is, that some groups of people are excluded when organisations reduce their usage of print-based forms of communication and rely more heavily on digital and social media communication because these are seen to be more cost-effective. Another important issue, noted by some commentators like Hearn (2013), is the effects that the spread of social media data mining across ordinary organisations has on the working life of their employees. The uptake of social media data mining by these organisations can be seen as a form of 'function creep' (Gregg 2011). Writing several years ago and more specifically about social media communications than analytics and insights, Gregg highlighted how the expansion of social media platforms led to a parallel expansion of tasks for ordinary workers in the knowledge economy, as they were increasingly required to maintain the social media profiles of their employers in addition to their existing workloads. As can be seen in the interviews discussed here, it is not so much social media profile maintenance but the analysis of social media activity and the conversion of this analysis into actionable insights that form today's function creep. Non-profit 2, for example, expects staff in all regional offices to assess social media insights, understand the causes of success and failure, and develop action plans for improved social media engagement. The journalists working in Media Organisation 2 need to convert insights, shared with them both live on the walls of their newsroom and retrospectively in the form of reports, into strategies for amplifying their audience numbers. Interviewee David, responsible for international, digital products, said that journalists receive a range of 'signals' from multiple analytics platforms, from which they are expected to develop an amplification strategy and ask themselves:

> What is my plan for amplifying what I've just done? How could I make that story sing? What headline is on it? Have I made sure it's been tweeted? Has it got the right hashtag on what I have tweeted about it? Have I got it tweeted by the right people? Because we have people like (a well-known anchor-woman) or some of the presenters who have tens of thousands of Twitter followers, if they tweet a story it's going to get read. How do we get them to endorse what we've just published?

Thus the roles undertaken by journalists and editors have multiplied; all are now expected to 'commission, edit, publish, build pages, amplify, curate', not just to edit and write stories. Journalists are still doing the work of journalists, but are also trying to incorporate these new analytics practices into their work, which is precisely what the organisation wants. David stated that it is 'every journalist's job, in my opinion, to make their story the best read they possibly can' and, according to him, additional knowledge derived from analytics makes it possible to achieve this aim.

Function creep has a deleterious effect on the quality of working life, as there is more work to be done, new skills to be acquired, and no more time in which to do either of these things. Writing about technological developments that have reshaped where and how office workers do their work, like mobile phones and networked computers, Gregg asserted that: 'The purported convenience of the technologies obscures the amount of additional work they demand' (2011, p. 2). The same could be said for analytics practices. Most interviewees extolled the virtues of social media data mining techniques and technologies, for themselves and for their organisations, and did not recognise the impact that the growing dependency on analytics has on the quantity of work that they and other employees are required to undertake.

Another parallel can be drawn with Gregg's argument that the extension of technologically mediated work leads to a 'constant onslaught of software and platform innovations, the bulk of which must be learned in employees' own time, (which) places the onus on individuals to keep step with the function creep affecting their jobs' (2011, p. 168). The rate at which social media insights tools proliferate is breathtaking—as noted earlier in this book, it is impossible to know all available tools at any given moment. Yet the employees we interviewed expect to manage their own understanding of this proliferation. The lack of organisational strategy with regard to analytics, discussed above, also means a lack of strategy vis-à-vis employee training and development in the use of these proliferating tools. The endless expansion of tools affects workers' capacities to predict and therefore manage their workloads, and means that there is a lack of clarity with regard to the limits of their job descriptions. All of this adds up to uncertain conditions for people working with social media data mining.

Furthermore, social media data mining has led to a requirement of more worker 'flexibility'. Some of our interviewees, such as Cathyrn in Non-profit 1 and Anthony in Council 1, noted that optimum times for social media engagement are outside standard working hours. Consequently, workers are required to work more flexible hours. But as commentators have noted in relation to other forms of knowledge work (for example Gill 2007, 2010), flexibility is not a mutual arrangement: the employer requires the employee to be flexible, but the employee cannot choose when he or she works. This represents a further deterioration in the quality of working life. Gregg has a term for this too: she uses the concept of 'presence bleed' to describe the dissolution of firm boundaries between personal and professional life that occur when the location and time of work become secondary to the requirement to get the work done. As Cathryn noted, in her organisation, social media engagement occurs at 7am, 7pm and at the weekend, and Anthony acknowledged that someone from his social media team would be 'listening' on bank holidays. The notion of presence bleed captures at least some of these changes.

Finally, social media data mining is serving as a form of worker surveillance in some of the organisations in which I carried out interviews. In Non-profit 2, Key Performance Indicators have been developed as a result of mining social media data, which are used to measure the effectiveness of their staff's social media engagement in regional offices. In Media Organisation 2, similar measures exist, with journalists expected to do their own self-surveillance with insights (asking themselves why did that story succeed and why did another fail?). As Hearn (2013) has noted, the implementation of social media analytics in the workplace may aim to help employees communicate, but it also allows employers to conduct 'deep employee monitoring'. Analytics claim to encourage better employee collaboration and innovation, but in reality subject workers to evaluation and assessment. This intensified management of workers with the aid of social media analytics is concerning, Hearn argues, because it enables perpetual evaluation and assessment of them. In contrast to techno-utopian discourses about the power and potential of new technologies at work, often equated with greater worker autonomy (Hearn 2013), social media analytics are used for increased monitoring, surveillance and discipline.

So, on the face of it, changes that result from social media data mining seem both mundane and minor, affecting the internal operations of organisations more than the services they provide in public domains. Some interviewees acknowledged that changes often relate simply to moving content around and making sure the right people get it. At the same time, such changes have significant effects on workers' lives. They represent a troubling increase in the monitoring of worker activity, and examples of both 'presence bleed' and 'function creep', even when they are undertaken with the purpose of providing better services to publics. This expanded use of data mining in organisational contexts also plays a role in the formation of particular attitudes to data and statistics, which I discuss in the next section.

Data Evangelism and 'The Fetishism of the 1000'

Most interviewees spoke evangelically about the power and potential of social media insights. This happened partly because, as noted above, it was easier to persuade evangelists to agree to be interviewed than doubters or critics. Some interviewees felt they needed to evangelise about analytics within their organisations in order to ensure internal buy-in and investment in data mining operations. Enthusiasm about what data mining can do and the numbers it produces is not surprising—social media and other forms of data mining generate interest in numbers because they generate numbers. Organisational involvement in social media data mining means that related data become available and this produces a desire for more and more numbers and statistics, as we saw in the previous two chapters.

In one example of data evangelism, when asked, Cathryn in Non-profit 1 said that she could not think of any limitations, issues or problems with social media data mining. She claimed that there is 'not too much one can know' and because of this belief, has never put any thought into potential problems or limitations. Similarly, Ruth in Council 2 listed numerous benefits and few disadvantages to doing data mining. She argued that as social media insights are easy to understand and often serve to explain phenomena, they are relevant to everyone, residents and council staff alike. Figures 6.1, 6.2, 6.3, 6.4, 6.5, and 6.6 show examples of the outputs of social media data mining taken from documents that interviewees shared

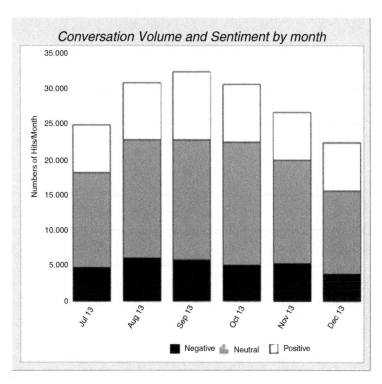

Figure 6.1 Some of the content from social media insights reports circulated within interviewees' organisations

with us, of the kind which Ruth describes as simple and self-explanatory. As can be seen, they prioritise quantitative data, such as likes and followers, conversation volume across platforms, numbers of retweets and shares, click-through rates, impressions, conversion rates, audience demographics and sentiment ratios. This focus on the quantitative produces an interest in and a desire for quantities, and this in turn produces the kind of evangelism for numbers, data and statistics witnessed in Cathryn and Ruth's observations.

And yet, although evangelical, interviewees also acknowledged the limitations of social media metrics. Technological limitations include tools' inability to capture comprehensive demographic or geolocation information; Ash in Media Organisation 1 said that traditional tools like Sysomos and Radian6 are 'blunt' because of this. Such criticisms suggest

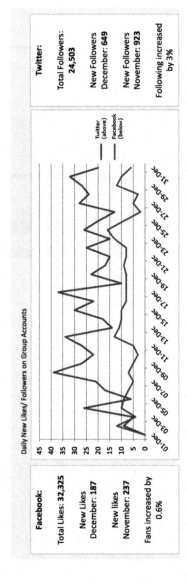

Figure 6.2 Some of the content from social media insights reports circulated within interviewees' organisations

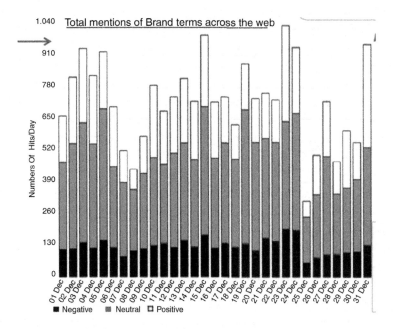

Figure 6.3 Some of the content from social media insights reports circulated within interviewees' organisations

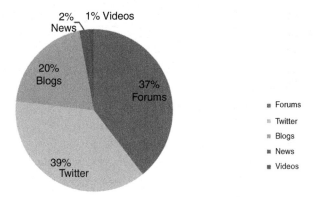

Figure 6.4 Some of the content from social media insights reports circulated within interviewees' organisations

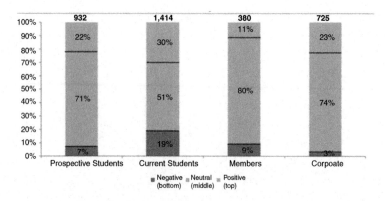

Figure 6.5 Some of the content from social media insights reports circulated within interviewees' organisations

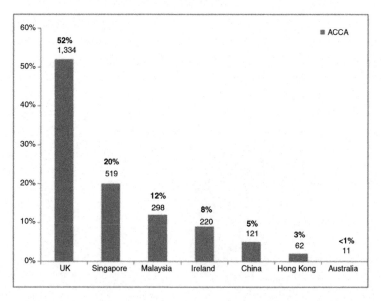

Figure 6.6 Some of the content from social media insights reports circulated within interviewees' organisations

that insights tools are not meeting users' expectations. Data also cause problems: there is too much, data are unstructured or badly formatted and therefore difficult to analyse, they are unreliable because users may have multiple accounts, or hard to compare because of constant changes in platforms' APIs and algorithms. Because of these technical issues, methodological problems emerge. Social media data miners need to be aware that measuring online is not equal to measuring audiences, as some interviewees claimed. The Dutch museums researchers said 'you never have all of the data; you're just capturing a moment in time' and Anthony at Council 1 said that social media insights are 'a finger in the wind'. Armed with these observations about the limitations of social media insights, interviewees sometimes bemoaned what they saw as their colleagues' uncritical belief in the power of data or their expectations that endless amounts of statistics could be produced. William in University 2 expressed his frustration with what he referred to as people's fascination with numbers, or 'the magic of numbers', which he called 'the fetishism of the 1000'. He claimed that within his organisation, there was a perception that the ability to cite numbers of people 'reached' through a campaign was proof in itself of a campaign's success. When a project has been completed, he said, if numbers can be produced—'that we've reached 50,000 and we've had 1000 people respond back to us about it, then that fulfills some kind of sense of requirement'. He felt that measurement was rarely undertaken with a genuine desire to self-evaluate, but rather was motivated by a desire to produce numbers, which were unproblematically equated with success.

Other interviewees lamented the fact that datafication produces certain problematic ways of thinking. Suzanne at Non-profit 2 gave an example. She said that if 18% of click-throughs on a certain post are from the UK and 10% are from Pakistan, there was a tendency to conclude that this provides concrete evidence that communities in the UK are more engaged and like the content that is being shared more than communities in Pakistan. In reality, she noted, these data may reflect different audience sizes in the two regions, or other factors, such as limited access to technology in the latter location. Like William above, others expressed concern that the very availability of numbers leads colleagues to want their own organisational numbers to be high. Ruth in Council 2 stated that, in a context where colleagues seemed 'blinded by stats and graphs', it was hard to persuade them that 100 fully engaged followers might be better than 1000 passive followers. Anthony in Council 1 told a similar story about a tweet which was retweeted three times, one of which reached such a large audience that there was little need for it to be retweeted further. In this case, he said, colleagues' view was 'Well it wasn't very good that, it didn't get many retweets.'

These patterns of evangelism, frustration when analytics do not meet expectations, and criticism of the evangelism of others might be seen as part of what Tkacz calls an 'emerging form of rationality' (2014), in which reliance upon data, statistics and numbers predominates. I have called this rationality a desire for numbers in previous chapters, bringing together Porter's (1995) ideas about *trust* in numbers with Grosser's (2014) argument that the quantification of sociality which results from social media metrics creates an incentive to increase likes, comments, friends and so on, which he describes as 'an insatiable desire for *more*' (2014, np). Within this rationality, argues Grosser, value becomes attached to quantification, worth is synonymous with quantity: if numbers are rising, worth is assured, and the desire for more is met.

Such desires can be seen in the comments of my interviewees, about their colleagues and their own views of social media metrics. Interviewees displayed a faith in what metrics can do, to paraphrase the title of Grosser's article, and sometimes demonstrated their own desire for more, such as David's desire for his journalists to amplify audiences in Media Organisation 2. Interviewees' observations on the limitations of insights technologies and methodologies can be understood in this context as frustration at metric practices not fulfilling their desire to produce numbers. Interviewees' comments on their colleagues' desire for more represents further evidence of this rationality, this belief that high numbers can be equated with worth and value. Cycles of data generation produce and reproduce an ever more prevalent desire for numbers. This notion of *desiring* helps us to make sense of interviewees' overstatement of the concrete changes that the numbers enable, their acknowledgement of the technical and methodological limitations of data mining and their frustration at their less insights-savvy colleagues' simplistic belief that big numbers are good numbers, as well as their own evangelism. As such, in times of datafication it is more fitting, I suggest, than the notion of *trusting* in numbers.

In datafied times, what was once qualitative is now measured quantitatively, so numbers are desired in relation to aspects of life previously the domain of the qualitative. This quantification of the qualitative should concern us because of what is lost when numbers are assigned such power, when numbers become cultural objects, and take on a new force. As noted in Chapter 4, Porter argues that numbers are for managing the world, not understanding it. In datafied times, when numbers become so widely desired, understanding is in danger of diminishing. Crawford (2013) alludes to this when she argues that big data (or big numbers) give us an overview of phenomena from a distance, but they do not enhance qualitative understanding. Because of these dangers, Baym (2013) argues that, 'Now, more than ever, we need qualitative sensibilities and methods to help us see what numbers cannot.' The powerful pull of numbers can also mean that some of the ethical issues associated with data mining and discussed earlier in the book are sidelined, as I noted in previous chapters and discuss in the next section.

'The Parasite on the Rhino'? Reflections on Ethical Issues

Because of my interest in thinking about what should concern us about social media data mining, I asked interviewees whether they felt that data mining raises ethical issues. After asking this broad question, as in the interviews discussed in the previous chapter, I described concerns relating to privacy, security and transparency with regard to social media data mining to them and asked them their views. On the whole, interviewees did not feel that their organisations' data mining practices raised ethical issues. For Cathryn in Non-profit 1, the only 'issue' was the lack of resources to enable her organisation to do *more* data mining. Suzanne in Non-profit 2 said that an issue for her was that people are increasingly aware of data mining and, as a result, are more guarded about their data, and that this undermines the accuracy of insights; this was seen by Suzanne as a barrier to data mining.

Responses to the concerns I put to interviewees relating to privacy, security and transparency were largely conventional. With regard to privacy, on the whole, respondents believed that as the data they analyse is in the public domain, it is acceptable for them to mine it, despite the fact that social media users might not see their data in the same way and have not shared them for these purposes. Unlike the professionals working in insights companies, who I discussed in the last chapter, these respondents did not acknowledge the ambiguity of concepts like public and private when it comes to social media. Participants generally did not have concerns about security, because they do not access personal data and they have appropriate data protection procedures in place. Jane from University 1 said 'the university is not watching you' and William from University 2 said that they are 'not infringing on data protection', they 'do nothing compromising' and are 'not identifying people'. When asked for their views on the proposition that social media data mining could be made more visible and that users could be made more aware of it, there was broad agreement that more transparency is a good idea, and that it would serve to educate social media users with regard to what happens to their data. Anthony in Council 1 said that there was a role for organisations to play in being more transparent about what they do with user data. However, there was also a strong view that transparency is the responsibility of the social media platforms, not organisations like their own which access social media data, either directly or through intermediaries. It was in reference to this point that Suzanne from Non-profit 2 used the phrase in the title of this section, 'the parasite on the rhino'. In this metaphor, social media platforms are the rhino, and organisations such as her own are the parasites, simply feeding off the data that the platforms accumulate and produce. In her view, the rhino is responsible for making data mining transparent, not the parasite.

Some interviewees felt that as their organisations provide services to the public, using social media data mining to help them achieve these objectives is ethically justified. Non-profits aim to serve the public or give people opportunities through education, and therefore are 'slightly different and special', according to Suzanne in Non-profit 2, compared to commercial companies. Or, as Anthony in Council 1 put it:

> If we can use this information to provide better services, more tailored services to the community, then the information that we are mining from them, despite our own, you know, some of us have strong privacy concerns about certain things, we have our own personal opinions ... But despite that, if it is providing better services and people have volunteered that information and we are not using it to target specific customers, but if we are effectively using it in—although it is not anonymised, we are using it in an anonymised method, then yes it is (ethical).

Here, despite justifying his organisation's data mining as ethically acceptable, in referencing his and his colleagues' 'strong privacy concerns', Anthony acknowledges that as an individual, he has ethical codes which may not be entirely commensurate with his employer's data mining practices. Similarly, some interviewees differentiated types of data mining as more or less ethical. For example, Cathryn in Non-profit 1, the member-owned family of businesses, identified some practices which she considered to be 'off-limits' for ethical reasons. These included sponsored blogging (the practice of paying bloggers to blog positively about particular products or brands) and setting up fake accounts to give the impression of more followers. She also differentiated between 'customer listening', which she felt was acceptable and 'colleague listening', about which she had some concerns. As a result of the PR crisis that this organisation had experienced when its CEO resigned in a haze of bad publicity, the organisation monitored what employees were saying about it on social media in an effort to manage their reputation. This long quote from Cathryn reveals her ethical discomfort with this strategy:

> There have been cases, like I've mentioned before, about employees saying things on social media which we've then picked up through the tool, and you know, they can have a disciplinary. Because they've mentioned people's names online. And things that are actually—they're actually illegal, when you go down to it. And I don't think they understood that it was. And in a way it does make me feel a little bit uncomfortable. Because when we've kind of noticed that, and we—the processes, if it's bad we have to send it to their line manager and they could be raised for a disciplinary. Which could obviously affect their job. That does feel a bit—it can feel a bit sneaky. Because obviously you can tell that when they've written it they didn't think that the (organisation was) going to read it. And we have found it through the tool. I suppose it can feel a bit sneaky. [...] Yeah, I suppose it can—it has made me feel a little bit uncomfortable in the past. Because you know that they didn't mean for us to see it. We have seen it, and they're going to get told off now. That's the bit, I think the colleague bit. The customer bit, of just listening to what customers are saying [...] It doesn't raise any issues for me personally about ethics, that.

It is interesting to note that, despite having said that she could see no issues or problems with social media data mining, here Cathryn identifies just that, a problem. As noted in the previous paragraphs, such contradictions could be read as what Gibson-Graham call 'ethical openings' (2006, p. 135), or things about which to be hopeful, as they reveal an openness to thinking beyond adopted ethical positions.

Jane in University 1 acknowledged that because of the type of organisation in which she works, there is a limit to the data mining that she can undertake: corporate actors can do extensive data mining, but as a university, she has to respect the privacy of students. Likewise, she acknowledged that social media users often do not read platforms' privacy statements and terms and conditions, and therefore she was one respondent who questioned the widely used argument that social media data are in the public domain and therefore 'fair game' for mining and analysis. She supported the view that social media users should be given the opportunity to consent to having their data mined. When discussing transparency in the interview with William in University 2, I suggested that organisations could make their data mining practices more transparent, so that their audiences are more aware of them and can choose to opt out if they wish. William replied that his team had considered publishing the reports that they produce from their data mining, so that their students and other interested parties know what they are doing and perhaps feel reassured about the ways in which data are being used. However, doing this might raise concerns about whether there is more mining taking place than is visible, he said. In the end, his organisation decided not to publish the reports because they include information which creates competitive advantage over other universities, which would be lost if they were to share their data. They did not want to give away too many of their secrets, William said.

When we asked interviewees about privacy, transparency and other ethical issues, some referred us to their terms and conditions and privacy statements, suggesting that these make clear what they do with data. Yet, in her book on online privacy and contextual integrity, Helen Nissenbaum cites a 2006 study which found that only 20% of website visitors read privacy policies. More recently, Marotta-Wurgler (2014) examined the privacy policies of 260 websites on which people share personal information, such as social networking sites, dating sites and message boards and found that, on the whole, they did not meet benchmark standards. For example, key terms (like 'affiliate' or 'third party') are rarely defined, and mitigation phrases and hedge words like ('from time to time', 'might', 'at our discretion') abound—she found an average of 20 per contract and a maximum of 53. She also found that policies change frequently, some 30 times a year. Thus the existence of privacy statements does not mean that they are clear, nor that reading them will lead to understanding of what happens to data shared on platforms and websites (I discuss what social media users in my focus groups said about such statements in Chapter 7). The privacy policies of the organisations discussed here include some clear statements, such as the first paragraph of Media Organisation 2's policy which states 'You consent to the data collection, use, disclosure and storage practices described in this privacy statement when you use any of our Sites (as described below), including when you access any content or videos.'[1] The statement goes on to describe the types of information

---

[1] The reference is not included, to preserve anonymity.

collected and how the organisation uses the information. The UK museum's privacy statement begins with a clear, bullet-pointed list of what it does with data, including what data are used for and when data are collected. But despite these efforts to be clear, these organisations' users are no more likely to visit relevant pages than users of other sites, and so they do not guarantee transparency about data mining practices.

Media Organisation 1 is an unusual example of an organisation which has chosen only to collect and analyse data voluntarily shared by registered users, who therefore might be assumed to be aware that they have consented to having their data mined. Believing that 'data was the new oil' as their CEO had declared, the organisation wanted to leverage the value of user data and so they developed a strategy which allows them to mine viewers' registration information, knowingly shared with them by viewers. To make this work, they felt they needed their viewers' trust, so they produced a video fronted by a popular comedian to communicate their intentions, which emphasises the benefits to viewers of data sharing and data mining, while at the same time, interviewee Ash stated, showing the organisation's commitment to transparency. In the 22 months that the scheme had been running at the time of the interview, they had accumulated over 10 million registered users. Ash said:

I think ultimately for brands to succeed longer term, they have to have and embrace a trust relationship with their viewers. Otherwise yes it's legal, yes it's all these great things, but if you're not starting that journey of trust now, you roll forward and I think very quickly consumers are going to be disenfranchised.

So this apparently ethical step of only mining data that is voluntarily shared by registered users is more than just ethical; it also provides financial advantage. As seen in the previous chapter, moral tactics also offer economic advantage.

CONCLUSION: CONSEQUENCES AND CONCERNS

The changes that result from the social media data mining discussed in this chapter appear extremely humdrum—chickens stocked further apart, changes to signage, quick responses to oil spillages. Often, the consequences of social media data mining relate to social media data mining itself, or to social media marketing. Targets are set, timings are changed, findings are shared, resources are re-directed and employees are set the challenge of amplifying audiences and striving for more—hits, tweets, shares, likes. As noted in Chapter 4, as we lower our sights to focus on the ordinary, we see actors in ordinary organisations lowering their sights, in terms of what they think data mining can enable, and what they want to do with it. The consequences of data mining cited here look so mundane it is hard to imagine that they might concern us.

But despite concrete changes seeming mundane and relating largely to internal, organisational operations, there are causes for concern in these ordinary and apparently humdrum examples. Adjustments made as a result of data mining may be small in scale, but this does not mean they are insignificant. Changes impact on the working life of employees in the organisations, and both 'function creep' and 'presence bleed' result from the introduction of social media data mining within organisations. The functions of staff in these organisations are expanding, as they are set targets like developing strategies to improve social media engagement and are expected to become data analysts themselves, in addition to existing roles. There is more work to be done, often outside conventional office working hours, because this is when people are engaging on social media—an example of the presence bleed that Gregg talks about. And there are more skills to be acquired, as tools, platforms and processes multiply, but no more time to do this in either. Social media data mining also provides organisations with extended capacities for worker monitoring, intensifying employee management and assessment. In these uses of social media data mining, we see evidence of a phenomenon that concerned some of the critics discussed in Chapter 3: its deployment for surveillance in spaces hitherto protected from such practices. In this sense, at least this criticism, out of those levelled at more spectacular forms of data mining, holds true in its more ordinary manifestations.

These changes that occur to workers' lives are imbricated with attitudes to data and statistics which emerge within metric, datafied culture and which I characterise as a desire for numbers. It is because of this desire that so much is expected of workers in relation to insights and analytics. Interviewees evangelising about the power of social insights, critiquing tools for not being able to meet expectations, or critiquing colleagues for not understanding these limitations all form part of an 'emerging form of rationality' (Tkacz 2014) in which a desire for numbers, seen as indicators of worth and value, predominates. Such commitment to the production of numbers and statistics can lead to an evasion of ethical reflection on issues relating to privacy, security and transparency, or at least to attitudes to these issues which do not acknowledge the nuanced complexity of these concepts in the context of social media. Interviewees had given these issues little thought, sometimes by their own admission. They spoke about them only when prompted, and then with limited recognition of the ethical dilemmas that social media data mining raises. Some interviewees thought that the public service remit of their organisations made their data mining ethically acceptable, at least more so than if they did it for the purposes of 'click chasing', as David in one of the media organisations put it. Some interviewees referred to their privacy statements as evidence that they are taking appropriate steps to inform users of their data mining practices. But some also noted practices that they considered off-limits, or ways in which data mining could be made more transparent to social media users. These small acts might be viewed hopefully as the opening up of a space for ethical reflection.

The small-scale organisational adjustments discussed in this chapter are important not only because they impact significantly on workers within the organisations, but also because of their cumulative effects. One cumulative consequence of the belief that with bigger numbers, worth is assured, this 'emerging form of rationality' in relation to numbers, is on the things that are lost when the quantitative alone is deemed to have value, like qualitative understanding. We might understand this desire for numbers as a new 'structure of feeling' (Williams 1961) (a term already riffed on in writing about data mining, as in the title of Hearn's (2010) article: 'Structuring feeling: Web 2.0, online ranking and rating, and the digital "reputation" economy'). Within this new structure of feeling, characterised by a desire for numbers, which in turn encapsulates evangelism, frustration, trust and distrust, there appears to be not much room for agency. Speaking up and being heard about the limits of data mining, or about its ethics, is not commonplace. In the next two chapters, I turn to other prospective 'agents' of data mining—social media users, data activists and social researchers—and reflect on their capacity to act with agency in the face of the harsh logic of these structures.

CHAPTER 7

Fair Game? User Evaluations of Social Media Data Mining

INTRODUCTION

In previous chapters, I have focused on the people doing, commissioning and experimenting with social media data mining in the contexts of ordinary organisations. One theme to emerge in these sites was a general belief in the idea that as social media data are public, they are 'fair game' to be mined, analysed and monitored by interested third parties. This perspective exists across the contexts discussed so far, even among professional insights workers who acknowledge the instability of the public/private distinction in relation to social media. The normalisation of this view is cause for some concern, I have suggested. Users' perspectives on social media data mining rarely figured in the discussions I have had with people mining user data. Users' views do not constitute evidence of whether these practices should concern us on their own, but they matter. They contribute to our understanding of the normative dimensions of social media data mining, not least because they provide insight into the thoughts and opinions of the people whose often intimate social media activity is considered by many analysts to be public data. So this chapter considers users' perspectives and reflects on how they contribute to addressing the questions at the centre of this book about what should concern us about social media data mining. In the chapter, I review studies by other researchers and I report on a series of focus groups I undertook in 2012 and 2013 with Dag Elgesem of the University of Bergen and Cristina Miguel of the

H. Kennedy, *Post, Mine, Repeat*,
DOI 10.1057/978-1-137-35398-6_7

University of Leeds. The research was reported in the article 'On fairness: user perspectives on social media data mining' (Kennedy et al. 2015a), and this chapter draws extensively on that article.

What users think about social media data mining has rarely been researched directly, though a number of quantitative surveys exploring attitudes towards digital data tracking cover related ground (such as Arnott 2012; Ball nd; Bartlett 2012; Lusoli et al. 2012; TNS Gallup 2012; Turow et al. 2005, 2009). The results of these studies are relevant to the subject of this chapter, especially because they provide diverse and not entirely consistent insights into user attitudes. I outline these diverse findings below. Qualitative researchers working with social media users to understand their usage strategies have contributed to understandings of users' perceptions of their social media activities, especially as these relate to privacy management. Important scholars in this area include boyd (2014), Marwick (2012) and Papacharissi (2010), among others. Their work is relevant too. Even though they have not focused specifically on attitudes to social media data mining, their insights into what users think they are doing as they navigate social media spaces help us to understand how users make sense of the data mining that the platforms undertake.

The small number of qualitative studies about users' attitudes to data mining which have been undertaken tend to mobilise one of the dominant paradigms discussed in Chapter 3, privacy and/or surveillance, to explore the extent to which users see social media data mining as constituting a form of privacy invasion or an extension of surveillance. In this chapter, I argue that focusing on privacy/surveillance in order to understand users' views is limiting in a number of ways. As I have noted, given the expansion of social media data mining, we need to think about whether criticisms developed in the context of its more spectacular manifestations help to explain its more ordinary forms. When it comes to thinking about users' perspectives, this means asking whether these dominant themes capture users' concerns. What if users do not think about their social media use in terms of privacy and surveillance? As I have noted, on social media, the absence of privacy and the presence of surveillance are serious problems, but users may not articulate their opinions about social media data mining using this language. So in order to advance understanding of how users think and feel about social media data mining, what is needed is a space in which they can express their views in their own terms, not least in order for researchers to be able to engage in conversation with users, in their terms, about social media and other forms of data mining. This is something that Andrew Sayer (2011) advocates when he argues that there is value in using lay terms in order to understand 'why things matter to people'—the title of the book in which he makes these arguments. So in the focus groups that I carried out with Elgesem and Miguel, we tried to move beyond the privacy/surveillance paradigm, by describing ordinary, real-world social media data mining practices to social media users, and asking participants to tell us what they thought of these practices in their own words.

There is something of a contradiction between the status of privacy in the framing of this chapter—and indeed, of the whole book—and in the previous empirical chapters. Here (and in Chapter 3) I suggest that we might need to move beyond privacy as a dominant framework for thinking about social media data mining, yet, in the last three chapters, I expressed concern about the limited recognition and discussion of the problematic character of privacy in social media contexts among research participants, which I described as an ethical issues. In those chapters, I attributed this absence to the power of the desire for numbers to subdue such debate. So, on the one hand, I am suggesting that privacy might not be important and, on the other, I am saying that it is. Both are true, I suggest—other issues surface as social media data mining becomes more and more ordinary, but at the same time, there are ethical questions which are not widely addressed by those engaging with social media data mining in ordinary organisations. Further contradictions with regard to privacy surface in the next chapter, in which I discuss movements which are pushing to open up data. Here, I report on research which addressed whether privacy matters to users, not by asking *if* privacy matters to them, but by asking them *what* matters to them in relation to social media and data mining. As we see below, privacy retains some significance for users.

In our focus groups, two themes emerged. First, talking to users about their views in a way which did not a priori frame social media data mining as relating to privacy or surveillance made it possible to understand the diversity in the findings of quantitative studies of attitudes to data tracking, referenced above and discussed in detail below. Separately, each of these studies suggests some patterns, but taken collectively, patterns break down, and are replaced by a mixed picture of diverse findings. This diversity can be explained, to some extent, through qualitative research which explores the thinking and feeling behind users' attitudes. Harper et al. (2014) argue that studies of attitudes to data monitoring fail to attend to diverse, individual and subjective responses to everyday tracking. Following them, in this chapter I argue that diverse attitudes to social media data mining relate to a range of factors, connected both to the individual and the data mining practice, such as: differing attitudes to whether social media content is public or private and to the need for consent and transparency; the type of data gathered and from whom; and the purpose for which social media data is mined. Thus I acknowledge the diversity of individual subjects, as Harper et al. suggest we should, but I go beyond this to recognise the diversity of social media data mining practices as well; here again, differences in data mining practices matter. This discussion also affirms Sayer's (2011) assertion that humans are fundamentally evaluative beings, a point to which I return later in the chapter.

Second, across this diversity and as a result of participants' reflective evaluations, a common concern about the fairness of social media data mining could be seen. Focus group participants tended to assess the fairness of specific practices and arrived at different conclusions about whether practices were fair or not, depending on their distinct evaluations of practices. I see this concern about the fairness of social media data mining as a form of contextual integrity in practice, to draw on the notion developed by Nissenbaum (2009), discussed in Chapter 3. Moving beyond the ethics of data flows to locate users' discussion of fairness in a context of broader debates, I suggest that it also relates to concerns about well-being and social justice. I conclude the chapter by arguing that social media data mining concerns users when it does not meet with their expectations of contextual integrity and when it disturbs their ideas about what is just and fair. But before I get to that, I provide an overview of studies of attitudes to data tracking and their findings, I describe and explain the methods used in focus groups, and I outline the diverse perspectives of our focus group participants on the fairness (or otherwise) of data mining practices.

WHAT DO USERS THINK? STUDIES OF USERS' VIEWS

Quantitative Studies of Attitudes to Digital Data Tracking

Quantitative studies of users' views on digital data tracking, which are relevant to my focus here, paint a mixed picture. Studies of attitudes to online tracking in digital advertising, such as those carried out in the US by Turow and others, conclude that people reject digital data tracking (Turow et al. 2005, 2009). In one study, 79% of 1500 adults agreed with the statement 'I am nervous about websites having information about me' (Turow et al. 2005). However, most other statistics from these studies point to ignorance rather than concern. For example, the 2005 report stated that around half of respondents did not know that websites are allowed to share information with affiliates (Turow et al. 2005), and the 2009 study found that 62% of respondents falsely believe that if a website has a privacy policy, it means that the site cannot share information about them with other companies (Turow et al. 2009).

A report produced for the European Commission into practices, attitudes and policy preferences in relation to the management of personal identity data (Lusoli et al. 2012) concluded that European citizens favour secure privacy and data protection rights. Based on a survey of more than 26,000 European Union (EU) citizens, the research found that social networking site (SNS) users are less cautious about sharing information about themselves than, for example, e-commerce users, even though they consider such information to be personal. Young people disclose more and control their privacy less than older users, but are equally worried about it, and they are significantly more comfortable with online profiling in exchange for free services than other user groups. The research also found that over half of SNS users have tried to change their privacy settings (a Pew Internet and American Life study (Madden 2012) found that similar numbers of respondents have done this), and around the same number claim either not to have been informed about or to be unhappy with the information available about the consequences of disclosing personal information on social media platforms. Framing their research in terms of the implications of the research findings for EU policy, the report's authors conclude that from a policy perspective, 'significant work is required to enforce informed consent and enhanced information about what may happen with people's personal data once it is disclosed' (Lusoli et al. 2012, p. 75). A similar EU-wide study involving around 2000 participants was undertaken by Ball and others (nd) to examine the extent to which EU citizens accept that infringements on online privacy are necessary in the name of security. Although precise findings are not available at the time of writing, the project website (http://surprise-project.eu/) suggests that participants want privacy legislation in relation to personal data to be strictly enforced.

In the UK, one of the three countries in which I conducted the focus groups, the think tank Demos produced *The Data Dialogue* (Bartlett 2012), a report based on a poll of 5000 adults' attitudes to personal information and data sharing. The research found that losing control of personal information was the biggest concern among respondents (80% were concerned about its use without permission; 76% about its loss). Measures which give the public more control were welcomed, such as being able to withdraw information (73%), or knowing what information is held (70%). As one respondent said, 'I'd really like it if websites and companies would tell me, in simple terms, what they know about me and what they do with it. I'd like the chance to control that information' (Bartlett 2012, p. 44). The report concludes that there is 'a growing crisis in consumer confidence' over the use of personal data, but no data is given to back up this claim. What's more, it is contradicted by a Pew Internet and American Life Project study of reputation management strategies (Madden and Smith 2010), which reports that, over time, users have become *less* likely to express concern about the amount of information available about them online—33% were concerned in 2010, down from 40% in 2006. The Demos report concludes that views about online sharing change when people are given control and choice about what data is shared and when the benefit of sharing data is made clear to them. It therefore suggests that consumers should be engaged in a dialogue about how data are collected and used, and be given meaningful choice and control over the information they share.

In Norway, another of the three countries in which we conducted focus groups, an online, panel-based survey of just over 1000 respondents (TNS Gallup 2012) showed that attitudes to targeted advertising are mixed in this country. While 36% of the respondents thought it was 'OK that Facebook serves ads on the basis of your interests', 54% thought this was 'not OK'. More people (71%) were negative about the idea of Facebook showing their interests to friends on Facebook. Only very few (3%) said they thought it was OK for Facebook to serve personalised advertisements based on their photos or on the content of their personal messages to friends.

In another example, a survey of public attitudes to measures to regulate the collection and use of online personal information in Australia, involving just under 1000 respondents (Arnott 2012), found that most respondents (75%) want more information about the way that websites collect and use information about people online. The research found that around a third of respondents were comfortable with tailored advertising and tailored news, with that figure rising to more than three-quarters if the tailoring is based only on information gathered from the website on which the tailoring is occurring, indicating a dislike of cross-site information sharing practices. The research found that almost all respondents welcomed all possible laws to regulate online data tracking which were suggested to them, including the right to privacy, do not track options, the protection of young people, and the right to know what information is held and request its deletion. This suggests widespread support for measures to regulate information gathering and sharing practices, an issue I return to later in the chapter.

The findings of these large-scale studies vary. Some studies identify concern; others tell a different story. In Turow et al.'s studies, data generally point to ignorance rather than concern. In the EU study, roughly equal numbers are and are not happy with the information they can access about what happens to their data. The Norwegian study found that three-quarters of respondents are not happy with Facebook sharing their interests with friends and so personalising their friends' feeds, but the Australian study found that three-quarters of its respondents accepted tailored content if it was not based on aggregated data. Pew studies have found that concern about personal data held online is diminishing, whereas the Demos study suggests that it is growing. This variation in responses is not explained in these quantitative studies, not surprisingly, as they aim to map the range and breadth of views, not to explain them. Writing specifically about surveillance studies' perspectives on public attitudes to information tracking, Harper et al. (2014) cite Lyon and Bauman, who claim that 'the public has enthusiastically or resignedly accepted such technologies' (2012, p. 4). Harper et al. argue that such macro-level, top-down assertions fail to attend to diverse, individual and subjective responses to everyday surveillance practices. They argue that 'what appears to be an enthusiastic or resigned acceptance of surveillance technologies may actually be much more complex' (Harper et al. 2014, p. 10). Following Wood (2005), they call for qualitative studies of 'the human dimensions' of technologies in context, which *acknowledge* variation rather than perceiving variation as a problem to be ignored. To explore what lies behind variation, what people are and are not concerned about, and whether differences in type, purpose and conditions of data mining practices inform users' views, qualitative enquiry is needed.

Qualitative Studies of Social Media User and Attitudes to Social Media Data Mining, and the 'Contextual Integrity' Framework

Many qualitative studies have explored users' perspectives on their social media usage and how these relate to strategies for managing social media privacy. These contribute to understanding the issues at the heart of this chapter. Early studies were often motivated by academic incomprehension of what motivated individuals to share intimate data so fully and so publicly on social media. More recently, studies have sought to contest the simplistic view expressed by Facebook CEO Mark Zuckerberg and others that, in a social media age, privacy no longer matters. Resulting studies with social media users have developed nuanced understandings of how notions of privacy, publicness and sociability become more complex in the context of social media. Some of these studies have been discussed in earlier chapters, such as boyd's extensive and often collaborative empirical research into young people's social media use, from which she concludes that privacy does indeed still matter, it just looks different in the context of social media (boyd 2010, 2014; boyd and Ellison 2007; Ellison and boyd 2013; Marwick and boyd 2010, 2014). Like boyd, other researchers have contributed to thinking about social media use, developing hybrid concepts like networked privacy (Marwick and boyd 2014), the networked self (Papacharissi 2010) and socially mediated publicness (Baym and boyd 2010) to point to changing relationships between the self, publicness and privacy. These and other empirical studies put users at the heart of studies of social media, as I do in this chapter. As such, they should inform our thinking about social media data mining, because they take users' views seriously, reminding us of why social media matter to users (to adapt the title of Sayer's (2011) book). At the same time, they develop understanding of social media usage in ways which do not construct users as passive and inactive in relation to data mining structures, instead highlighting the diverse, sophisticated and sometimes agentic ways in which users navigate social media spaces.

The small number of qualitative studies that have focused specifically on attitudes to data mining are generally concerned with 'the privacy paradox', as discussed in Chapter 3. That is, users' sharing practices appear to contradict their expressed privacy concerns. As noted above, some authors researching this phenomenon conclude that a distinction between social privacy (controlling which people within users' networks get access to their information) and institutional privacy (the mining of personal information by social media platforms and other commercial companies) is needed. Raynes-Goldie (2010), for example, argues that social media users *do* care about controlling their personal information, but that their concerns relate to social privacy, not institutional privacy. Young and Quan-Haase (2013) build on this distinction in their study of privacy protection strategies among students, in which they argue that an absence of institutional privacy and a related presence of data mining practices have become social norms. The researchers asked their participants about 'their general concern for online privacy and privacy on Facebook' (2013, p. 479) and about what measures they have adopted to protect their privacy. Respondents answered by talking about 'social privacy' measures, from which the researchers conclude that people do not care about institutional privacy. But we do not know whether respondents were aware of the types of practices that might be considered institutional privacy breaches, or if they were aware of measures they might take to protect their institutional privacy (such as advertisement blocking). It is possible that, even where such awareness exists, social media users do not see social media data mining as a privacy issue, so when asked what privacy issues concern them, they do not talk about it. Here we can see that adopting a specific privacy lens may inhibit discussion of users' more significant concerns, of what matters to them.

Similar problems can be identified in research about social media and surveillance. For example, Humphreys (2011) asserts that participants in her study of the check-in app Dodgeball understand and care about social, not institutional, surveillance. Like Young and Quan-Haase, she bases this assertion on the fact that none of her respondents mentioned or expressed concern about what she calls 'corporate or state surveillance', which again may be more indicative of a lack of awareness than a lack of concern. In these examples, it is unclear whether research participants were aware of the practices about which, it is claimed, they are not concerned. More detail about methods is needed in order to ascertain this. What *is* clear is that studies which aim to produce findings about privacy or surveillance generally produce findings about privacy or surveillance and, in doing so, they do not appear to have created spaces in which participants can talk about what matters to them in relation to social media data mining.

To address this limitation, in our focus group research we were interested in exploring what social media users think about social media data mining *beyond* privacy and surveillance. We designed our research to enable participants to express their concerns, if indeed they had any, in their own terms and to tell us, again in their own terms, what mattered to them. We did not mention privacy and surveillance. As I show below, we found that the fairness of social media data mining practices was a concern for our participants. Nissenbaum's (2009) concept of 'contextual integrity' is a useful tool for thinking about this concern with fairness, although, as Sayer notes with regard to his own term 'lay normativity', this is 'a rather alienated way of describing things' (2011, p. 2). Nissenbaum argues that, when people disclose personal information in a particular context, they have particular expectations of what will happen to their data within that context. Nissenbaum writes:

> finely calibrated systems of social norms, or rules, govern the flow of personal information in distinct social contexts (for example education, health care, and politics). These norms, which I call context-relative informational norms, define and sustain essential activities and key relationships and interests, protect people and groups against harm, and balance the distribution of power. Responsive to historical, cultural, and even geographic contingencies, informational norms evolve over time in distinct patterns from society to society. Information technologies alarm us when they flout these informational norms—when [...] they violate contextual integrity. (2009, p. 3)

People care about 'appropriate flows' of information, asserts Nissenbaum, not just about control and secrecy, and ignoring these is a violation of people's data rights. What is needed, then, is contextual integrity, or respecting people's expectations of what might reasonably happen to their data within a given context.

Nissenbaum defines contextual integrity in terms of informational norms that 'are specifically concerned with the flow of personal information—transmission, communication, transfer, distribution, and dissemination—from one party to another, or to others' (2009, p. 140). Informational norms take account of the capacities or roles of particular actors, the types of information, and the principles under which this information is transmitted among the parties. Such norms distribute rights (such as the right to consent) and responsibilities (such as the duty of confidentiality) among the actors involved in the communication of personal information and data in a particular context. These are the basis on which people form expectations of how data will be managed and processed. As I show below, the social media users in our focus groups were uncertain about what norms should apply to social media data mining and differed in their views about which aspects of particular practices are morally troubling. I characterise their talk about these matters as demonstrating a concern about the fairness—or otherwise—of social media data mining. Different attitudes to social media data mining resulted from different evaluations of whether practices were fair or not. This, I argue, is contextual integrity in practice: case-by-case assessments of whether data mining practices can be considered to be reasonable, fair and ethical. I develop this point below, but first I describe our focus group research.

FOCUS GROUP METHODS FOR RESEARCHING USER PERSPECTIVES

Elgesem, Miguel and I carried out ten focus groups with 65 participants across three European countries (England, Norway and Spain) to explore what social media users think about social media data mining. We wanted to produce knowledge that was not specific to one country, and, as one of us is English, another Norwegian and another Spanish, it made sense to carry out the focus groups in these countries. We chose focus groups because they allow access to a relatively large amount of views in a short period of time; they give access to attitudes, feelings, beliefs and reactions, which is what we wanted to explore; data can be produced through interaction; and participants may take the initiative in the discussion (Gibbs 1997), something we felt would enable them to express their views in their own terms. We used a snowballing method to recruit participants who were often therefore known to each other. We opted for some homogeneity within focus groups, not only because this made it easier to recruit participants with the limited resources we had available, but also because we subscribe to the principle that homogeneity results in understanding of others' lifestyles and situations and so facilitates discussion (Krueger and Casey 2008). We aimed to recruit participants who represented a cross-section of social media users, acknowledging that in this qualitative study, our groups would not constitute a representative or statistically significant sample. We succeeded in recruiting participants of various ages, educational backgrounds and work experiences, but there was less racial and ethnic diversity among participants (almost all of whom were white) than we would have liked. We characterised our groups as: teen bloggers; young people with mental health difficulties and their support workers; undergraduate students; Master's students; low-income users (who earn less than £950 ($1350) per month per month); marketing professionals (who might use social media data mining techniques); mothers in their 30s, who work part-time; professionals in their 40s; older users (aged 56–83). Groups and their membership are summarised in Table 7.1.

There were 19 male and 46 female focus group participants; we found it easier to persuade women to participate than men. The youngest participant was 15 and the oldest was 83, with people in their 20s best represented. Prior to the focus groups, participants were asked to categorise their social media use: 21 described themselves as avid users, 29 as regulars users and 15 as occasional users. In addition, participants were asked to list the social media platforms that they use. All participants used Facebook, except one British professional in her 40s, who only used Twitter and LinkedIn; 35 used Twitter; 18 used Instagram (the majority of whom were from Norway); 16 had a blog; 15 used LinkedIn; 9 used YouTube; and 8 used Pinterest. Other SNSs mentioned included Flickr, Tumblr, Tuenti, FourSquare, CouchSurfing, MySpace and Badoo.

Within the focus groups, we shared factual statements about real-world social media data mining activities with our participants, which we called scenarios. We did this for a number of reasons. First, it was clear from other studies that we could not assume that people know about social media data

Table 7.1   Focus groups and members

Group	Country	Number of participants
Teen bloggers	Norway	8
Master's students	Norway	8
Professionals in their 40s	Norway	8
Young people with mental health difficulties and their support workers	UK	6
Undergraduate students	UK	8
Mothers in their 30s, who work part-time	UK	6
Professionals in their 40s	UK	5
Older users (aged 56–83)	UK	6
Low-income users (who earn less than €1200 per month)	Spain	5
Marketing professionals (who might use social media data mining techniques)	Spain	5

mining—we needed to tell them about it through examples. Second, we were interested in what people think about *actual* practices, rather than possible or imagined practices, following McStay's (2011) warning, cited in the previous chapter, about the dangers of confusing the actual with the potential. Third, acknowledging that not all social media data mining practices are the same, we wanted to explore whether differences among them—for example, type of information gathered, purpose, by whom, in what context—had a bearing on attitudes. We wanted to find out whether practices concern people and if they do, which ones. We worked with commercial and academic social media analysts to develop the scenarios. We wanted them to be representative of ordinary social media data mining practices, in contrast to the more spectacular forms and in keeping with this book's focus, but we also included examples of Facebook's data mining, as we felt that talking about a familiar platform would help participants make sense of other, perhaps less familiar practices. We asked participants to note their immediate responses to the scenarios on a sliding scale, from 'this concerns me' then 'this is not OK' at the concerned end, via a neutral position of 'I have no opinion' in the middle, to 'this is OK' and 'this is a good thing' at the unconcerned end. Scenarios and a summary of responses are included in Table 7.1 below. These responses do not represent the participants' considered judgements and they may well have modified their opinions in the later discussions in which we reflected on these initial, written responses. Three sets of scenarios were presented to participants, covering data mining practices undertaken by (a) the social media platform Facebook, (b) intermediary social media data mining companies and (c) academics. I describe these in more detail in the next section, where I also outline our findings and highlight emergent issues.

WHAT DO USERS THINK? FOCUS GROUP FINDINGS

Diverse Perspectives

The first set of scenarios (1a–g in Table 7.2) focused on Facebook's data mining activities and were taken largely from Facebook's own Terms and Conditions. Responses to this set of scenarios varied. Just under half of the participants did not like Facebook's practice of using personal information for targeted advertising (ticking either 'This concerns me' or 'This is not OK') with others responding positively to this practice ('This is OK' or 'This is a good thing') or indifferently ('I have no opinion'). However,

Table 7.2   Scenarios and participants' initial responses

	'This concerns me' + 'This is not OK'	'I have no opinion'	'This is OK' + 'This is a good thing'
1. Facebook			
(a) Facebook analyses your activities on Facebook and the information that you provide to Facebook in order to target advertising to you.	47%	17%	36%
(b) Some of the information Facebook uses to show you targeted ads includes: things you share with Facebook (e.g. liking a page); information you put on your timeline (e.g. your gender; information from third parties).	53%	17%	30%
(c) Facebook shares the data it gathers about you with third parties. Facebook works with others to combine information they have collected from you with information from elsewhere, to enable the advertiser to send you targeted advertising on Facebook.	76%	4%	20%
(d) Facebook shares data anonymously.	14%	18%	68%
(e) Advertisers who access this anonymous data may be able to identify you by putting the Facebook data together with data from other parties. They can then track individual users' activities across online spaces.	86%	9%	5%
(f) Facebook shares data that you have chosen not to make public in your profile. It ignores privacy settings when it comes to sharing data with advertisers.	98%	0%	2%
(g) Facebook does all of this with cookies, small files that are dropped on your computer when you're using Facebook. These cookies track your online activities both on and off Facebook, with the aim of delivering ads that are relevant to you.	62%	16%	22%
Total	**60%**	**12%**	**28%**
2. Social media data mining companies			
(a) Social media data mining companies collect information from blogs, Twitter, public accounts on Facebook, discussion forums, to see how brands, products or organisations are talked about.	26%	22%	52%
(b) They collect this information without users' consent.	81%	3%	16%

(c) A multinational company DAPS (which includes a global chain of coffee houses) wanted to know what its employees were saying about it in social media. It believes that its employees are part of the product that it offers, because it is a service industry, so it employed a social media data mining company to find out about this.	78%	5%	17%
(d) The social media data mining company found groups of people on LinkedIn, Facebook and Twitter who identified themselves as DAPS employees, and posts to forums about work issues where the posters identified themselves as DAPS employees. It analysed what people said about working at DAPS.	69%	9%	22%
(e) DAPS might use this information to plan HR and PR strategies, such as devising a policy relating to how DAPS employees who identify as such use social media, providing social media usage training to employees, or changing how they market and recruit.	40%	20%	40%
(f) A mother-and-baby product retailer wanted to understand the concerns of first-time mothers, so it could include content that was right for them on its website, and so direct them to buy its products. So it employed a social media data mining company to find out.	35%	17%	48%
(g) The social media data mining company analysed posts on forums like mumsnet.com, where some new mums post comments that they know will get noticed by brands, because the forums are high profile.	14%	29%	57%
(h) Other new mums also share very personal posts, for example about difficulties and pain related to breastfeeding, sleep deprivation, fears about their partners' infidelity.	38%	19%	43%
(i) The retailer used the information that resulted to try to attract new mums to its website.	31%	20%	49%
Total	**46%**	**17%**	**37%**
3. Blogs and MySpace			
(a) A group of academic researchers analysed the way that young people (aged 13–24) share their suicidal thoughts on MySpace (when MySpace was more popular than Facebook), thoughts that they might not be comfortable sharing in person.	34%	11%	55%
(b) Some comments were addressed directly to face-to-face friends.	50%	12%	38%
(c) Some comments were identified as cries for help.	36%	16%	48%

(continued)

Table 7.2 (continued)

	'This concerns me' + 'This is not OK'	'I have no opinion'	'This is OK' + 'This is a good thing'
(d) Permission to do this was not sought or received from the young people. They didn't know it was happening.	71%	12%	17%
(e) A number of comments were believed to be written by people with a serious potential of committing suicide. No action was taken to follow up on identified suicidal thoughts.	89%	9%	2%
(f) The people carrying out the research felt that the findings could be used to inform health professionals about the ways in which adolescents communicate suicidal thoughts in social media.	13%	14%	73%
(g) Another group of academic researchers collect blog posts about climate change, gathering information such as the name of the blogger, links to the post and from the post, and the content of the posts.	20%	9%	71%
(h) They are doing this to help them build a tool which analyses how information is spread in blogs.	10%	23%	67%
(i) The aim of the tool is to make information available on a much wider scale; intended users include researchers, media monitoring companies and the general public.	18%	24%	58%
Total	**38%**	**17%**	**45%**

many participants were negative about the fact that personal information is shared with third parties, and the majority was worried that, by combining Facebook data with information from other sources, their activities could be tracked across online spaces. Almost all of the participants responded negatively when told that Facebook ignores users' privacy settings when sharing information with third parties.

The second set of scenarios (2a–i) related to the activities of commercial social media data mining companies, which mine social media data on behalf of clients. These were drawn up in collaboration with professionals working in such companies. Examples included the use of social media data mining by a fictitious multinational company to find out what its employees were saying on social media platforms about working for it, and by a mother-and-baby product retailer that wanted to find out what concerns first-time mothers, so it could write relevant content for its website and so attract customers. Half of the participants had no problem with these kinds of social media data mining and the rest were either indifferent or responded negatively. But the majority (81%) responded negatively to the fact that information is collected without consent. A similarly large proportion indicated that they felt negatively about the fictitious company's monitoring of what its employees say about the company on various social media sites. Fewer participants were sceptical about the mother-and-baby product retailer example than to the previous one. More than half of the participants responded positively to the description of the forum monitoring. A few (14%) were negative about this monitoring and some (29%) were indifferent. Some of our participants switched from feeling indifferent to feeling negative when they were told that the information collected could be described as sensitive.

The final set of scenarios (3a–i) focused on academic uses of social media data mining. It included a study of the way that young people share their suicidal thoughts on MySpace and another study of blog content about climate change (in which Elgesem was involved). These statements were drawn up in collaboration with the academics whose research was cited. Half of our participants were generally positive about the monitoring of young people's conversations about suicide but some (34%) were negative. A majority found it problematic that the young people were not informed about the activity and 89% were critical of the fact that the researchers did not take any action to follow up on information about users who were deemed to be at risk of committing suicide. Two-thirds of our participants thought that the collection of data from blogs about climate change was acceptable. However, some became more sceptical of the activity when told that the tool that the researchers hoped to develop as a result of the study would make information in blogs available for further analysis on a much wider scale.

Thus participants' initial responses to the scenarios were diverse. There were more negative than positive responses, more concern than indifference or support for the social media data mining practices we described. Concern focused on issues of consent, transparency and respect for the contextual integrity of shared data, for example Facebook's sharing of data designated private by users. So, how did participants account for their initial responses in the discussion that followed, and to what extent was this also characterised by diversity? In the discussions, social media data mining was characterised as disconcerting, disgusting, unpleasant, frightening, worrying, scary, annoying, sinister and intrusive. Some participants were surprised to hear about the potential consequences of cross-platform tracking and information sharing, and expressed concern about the ways in which information can be pieced together. For example:

> By finding your patterns and what your preferences are, they can find so many things about you that you don't want people to know. (male, 44, Professionals group, UK, IT analyst)

At the same time, some of our participants were not concerned about the practices described to them, and noted the benefits of social media data mining. One said that data mining is a good thing 'because it makes advertising relevant' (female, 30, Marketing professionals group, Spain, community manager). Another stated:

> It can work to your advantage, can't it, getting offers, information that you need, it can be a good thing. (female, 38, Mothers group, UK, psychiatric nurse)

Such variation could also be seen in relation to discussion of the specific scenarios. None of the scenarios met with consistent, consensual responses. The scenario which raised least concern was the use of social media data mining by academics to study how information about climate change is shared on blogs. However, some of the teen bloggers expressed concern about this study, seeing their own blogs as more private and intimate than other participants understood blogs to be, and aware that some bloggers could be young or vulnerable. As the youngest participant put it, 'I just thought it's a bit different if the bloggers are very young' (female, 15, Teen bloggers group, Norway, school student). Some participants were concerned about the study of expressions of suicidal thoughts on MySpace, because of the vulnerability of the subjects under investigation and the seriousness of the subject matter. But some were supportive of it, because it could be for the social good. Others wanted to know the precise outcomes of the study before they would comment—were the young people helped? Similar patterns could be seen in relation to the other scenarios. The example of the multinational company mining its employees' social media conversations met with a range of responses: some respondents felt that the company had a right to know what its employees were saying about it in networked public spaces; some felt the findings could be used for the good; others felt this was intrusive.

Participants' diverse responses to the scenarios confirm Harper et al.'s assertion that responses to data tracking often vary. In their view, this variation arises from the ways in which human subjects differ from each other. In the case of our participants, individual differences such as age, nationality, occupation, social media use and awareness of similar practices seemed to inform their perspectives, as younger participants tended to be less concerned about monitoring scenarios than older participants, UK participants were less concerned about commercial company and academic monitoring than their Norwegian counterparts, Norwegian professionals expressed more concern about all scenarios than the younger Norwegians, and marketing professionals appeared less worried than all other groups in relation to all scenarios. While a causal relationship between individual differences and attitudes to social media data mining cannot be established categorically, in some cases participants stated explicitly that factors such as occupation influenced their attitudes. One participant from the UK professionals group who worked in IT said that his work made him aware of and concerned about data mining. Most marketing professionals claimed that their work meant that they were unconcerned by some of the scenarios. One said:

> I know how everything works, so I think it's fine, it's always to give a better service to the user, I'm perfectly fine with everything. (female, 30, Marketing professionals group, Spain, community manager)

Participants' prior awareness of the techniques of digital advertising also seemed to inform their perspectives on social media data mining, as in general, the greater their awareness of digital advertising, the less their concern. One unconcerned participant said:

> Isn't this just a progression in advertising? Advertising has always done this. (female, 36, Mothers group, UK, part-time teacher)

Similarly, participants who see social media as public platforms thought that data mining was acceptable. Some participants asserted that social media platforms are public spaces and should be seen as such by their users (thus subscribing to the 'it's public so it's fair game' ethos identified in previous chapters), whereas others acknowledged that, despite this, social media platforms sometimes *feel* private, and others differentiated platforms on a private–public spectrum. Thus participants' perceptions of the privacy or publicness of social media seemed to be another factor that influenced their perspectives on social media data mining. Talking about social media data mining professionals, one respondent who was not concerned by the scenarios stated 'these people have just tapped into a public domain' (female, 22, Undergraduate students group, UK, sociology and social policy student). Similarly, a number of users appeared to believe that the trade-off between connectivity and being tracked was a fair one. However, some were concerned to hear that Facebook shares data that its users designate as private. These social media users expected a degree of privacy when using this platform and were concerned about the mining of material they had designated as private:

> If I've set it to private I expect it to be private. I don't expect it to be subjected to what Facebook thinks okay to give advertisers or not. (female, 20, Undergraduate students group, UK, television student)

Others demonstrated sophisticated understanding of the distinction between *being* private and *feeling* private in social media spaces (a distinction that I have written about in relation to historical new media forms (Kennedy 2006) and which I return to in the conclusion of this chapter). For them, feeling private but being public was a problem. One said:

> It feels like it's a safe way of communicating to lots of people, it's set up like that, there's all this confidentiality, you can avoid sharing it with everyone and you can keep it private—but can you really? (female, 47, Professionals group, UK, consultant psychiatrist)

Participants sometimes differentiated social media from other media, identifying them as personal and intimate spaces. In the group of older users in the UK, most participants disagreed with one group member's assertion that targeted advertising on social media is like TV advertising aimed at the types of people assumed to be watching during particular schedule slots. The other group members stated that these are not comparable, precisely because of the intimacy of social media, with one stating 'on the computer it's almost a more personalised environment than watching the television' (female, 83, retired teacher). In other groups, some participants indicated that their expectations varied depending on the platform. One said:

> If it is on Facebook, yes [I am concerned]. If it is a forum, I don't care. But a social network is something more personal. (male, 32, Low-income group, Spain, unemployed)

For these users, some platforms are more personal than others—SNSs are considered to be more intimate than forums, for example. Throughout all of these conversations, I suggest that participants were consistently assessing whether data mining practices are fair. This particular participant suggested that it is fair to monitor forums, because they are open and public, but not SNSs, because they feel more personal and intimate. For others, social media data mining is always fair, whether the spaces and content feel intimate and personal or not. Considering the fairness of data mining practices was a common occurrence in participants' responses to scenarios. In the next section, I say more about this.

Common Threads: Concerns About Fairness

In this section, I develop the proposal that a concern about fairness was a common factor in participants' responses to data mining practices. I focus on the characteristics of specific data mining practices which appeared to influence participants' assessments of them as fair or otherwise, including the type of data gathered and from whom, and the uses to which mined data are put. Returning to Nissenbaum's concept of contextual integrity, these can be understood as context-relative considerations which inform expectations about data flows. Her discussion of the capacities and roles of individuals, the types of information and the principles under which information is transmitted map neatly onto the issues that our focus group participants raised, in which the principles of consent and transparency played a particularly prominent role. I propose that this concern about fairness can be understood as contextual integrity in practice. Not just an interest in individual privacy, this represents a broader concern with whether social media data mining practices contribute to well-being in social life.

Some participants differentiated the types of social media data that they felt it was acceptable to monitor. There was greater acceptance of the monitoring of information shared with Facebook when setting up an account, or liking something, compared to personal or intimate posts, as expressed in this quote:

> If they only take the information you register when you join Facebook and the pages that you like, I think it's okay to get ads directed to me based on that. (female, 15, Teen bloggers group, Norway, school student)

Various comments of this kind were made across the groups, in which monitoring likes was seen as acceptable, but monitoring personal or intimate posts was not. Thus participants made distinctions regarding the source of monitored information. In addition, the type of information gathered was a factor: 'it depends on what that information is for me as to how I'd answer that question' (female, 38, Mothers group, UK, civil servant) said one participant when responding to one of the scenarios we described. Some participants were also concerned about who the data came from, with concern about monitoring children's data sometimes expressed. This was not just a concern about the collection of their data, but also about children's ability to resist the personalised and targeted messages that they might subsequently receive. This is what Nissenbaum describes as a concern about the capacities of individuals.

I argue that a concern about fairness characterises these responses. Respondents thought that it was fair for Facebook to use data that they had given to Facebook for its own purposes, but not to use data that they 'give' to their friends, for example through private messages or personal wall posts. Likewise, respondents thought that it was unfair for Facebook to take data from vulnerable individuals like children and to target advertisements at them. A concern about fairness also influenced participants' comments on how the purposes of data mining activities informed their perspectives. As one participant said:

> Collecting the information isn't the worrying thing for me, it's what happens with it. (female, 36, Mothers group, UK, self-employed)

There were many similar statements. Some support was expressed for data mining for security purposes, as this was thought by some to be in the public interest and therefore fair, although others acknowledged that historically, such data mining has not always been done well or fairly. Some participants were not concerned about mining for advertising purposes, because this was seen as an exchange, and therefore a 'fair' game—targeted ads are received in exchange for a free networking service. Thus the uses to which mined data are put intersect with participants' individual ethical barometers, which inform their expectations of the principles under which data should flow. For some, mining for targeted advertising is acceptable, a fair exchange, but for others, it is not, because it is intrusive. This point was captured nicely in this mildly self-mocking comment from one participant:

> If they do something with it that I agree with, then that's okay; if they're going to do something I don't agree with, then it's not. (female, 49, Professionals group, UK, welfare rights officer)

Although humorous, the above quote is important in that it shows how values influence user evaluations of data mining practices, as Sayer (2011) suggests they do in relation to broader social practices. Or, as another participant suggested, it is possible for mining to take place for the good, or its opposite:

> If it's going to be used for benefit, if they find a trend on young people committing suicide and that can be avoided, that's fine. But when they can find a trend of what we're going to do next week, that's not fine. (male, 44, Professionals group, UK, IT Analyst)

The final principle that influenced participants' perspectives is consent, whether users agree to the monitoring of their social media activity and, relatedly, whether the existence of monitoring is sufficiently transparent to users—that is, whether users are aware that they are consenting at all. One participant asked if the scenarios involved monitoring with or without users' consent, and when told that all were without consent, stated 'that's my sticking point'. For many participants, this was a problem, even those who were generally unconcerned about the data mining activities described to them, and this was evident in participants' initial responses, as well as their subsequent discussion. Many participants felt that even when social media platforms communicate to their users about what happens to their data, such information is hard to understand, often written in 'five billion words', as one participant put it, or hidden away in fine print, as another stated. This discussion captured participants' lay perspectives on some of the problems identified by Marotta-Wurgler's study (2014), referenced in the previous chapter. There was widespread belief that because of these difficulties, users frequently do not read the Terms and Conditions to which they agree on social network sites.

According to some participants, because of the unclear ways in which information about data mining is communicated to social media users, there is a lack of awareness about such practices:

> They are collecting my data without me knowing. It would be another thing if I was to give them the data. (female, 30, Low-income group, Spain, unemployed)

However, this view was not shared by all. Some participants believed that the platforms communicate what they do with users' data and so they have acted fairly. It is then the user's responsibility to understand what s/he is consenting to:

> If you do not know the network you are using then I don't think you have the right to complain about the way they share the data [...]. If you have not bothered to find out what the rules are, you cannot complain about how they share your information. (female, 15, Teen bloggers group, Norway, school student)

Such views were in the minority. As a result of more common concerns about the absence of informed consent, a number of participants advocated greater transparency about social media data mining. Some participants proposed that simpler, clearer, more transparent information is needed:

> Perhaps there should be a requirement from the government that a warning should be issued about the risk that your information could be used if you have a public profile. (male, 27, Masters students group, Norway, engineering student)

Two participants in different groups used the metaphor of a cigarette packet to describe ways in which greater transparency could be achieved, proposing that social media platforms should include 'a big banner at the bottom, like "smoking kills" on a cigarette packet' (female, 49, Professionals group, UK, welfare rights officer). These participants wanted simple, transparent information about social media data mining. Thus, although we encountered diverse views in our discussions of specific social media data mining scenarios, which could be interpreted as uncertainty about what norms should govern the mining of social media data (as also seen among commercial social media data miners discussed in Chapter 5), one common thread was a desire for more transparent information about what happens to data shared on social media. This in turn can also be understood as a desire for fairer social media data mining practices than currently exist.

In these examples, participants evaluated the fairness of the data mining practices presented to them, and they came to different conclusions. These differences result from users' differing evaluations of whether a monitoring practice is fair or not. A fair practice is one that is consistent with a user's expectations of contextual integrity concerning the collection and use of social media data in particular contexts; different attitudes to social media data mining result from different evaluations of practices. For example, several participants argued that Facebook's use of data for the purpose of selling personalised advertisements is reasonable because this makes the service possible. It is a fair deal, they suggested. However, for many, it is still considered necessary to inform users of these practices, as only then are the practices consistent with the user's expectations of contextual integrity. Contextual integrity involves considering the transmission principles for particular practices, for example, who should have access to data, what obligations recipients have to keep it confidential, and what type of consent to its processing is required. The norms regulating transmission principles are to a large extent a function of the attributes of the data processing, argues Nissenbaum, such as the purpose of the processing, the intended audience, the sensitivity of the data, whether it is meant to be persistent or ephemeral. The vulnerability of subjects is also an important factor that influences evaluations of what principles should govern the transmission of personal information. All of these attributes surfaced in participants' discussions, in which they questioned how these data attributes map onto transmission principle norms in different social media contexts—although of course they did not express their concerns in these terms. In 2012, Kirsten Martin carried out a study of user engagement with privacy policies in which she asked participants to read privacy policies, then posed scenarios to them, asking (a) is this problematic, and (b) is this within the realms of the policies. Participants all answered yes to (a) and no to (b), even though all scenarios were all within the policies' realms. Her research suggests that users do not simply misunderstand such policies; rather, they have substantive expectations about what a privacy policy should do (Martin 2012). Such expectations are integral to contextual integrity, and we saw them in operation in the focus group participants' discussions too, as they made case-by-case assessments of whether data mining practices are characterised by appropriate, context-relative norms and therefore can be considered to be reasonable.

This concern about the fairness of data mining practices moves beyond an interest in one's own individual privacy, to a broader interest in whether social media platforms are operating within normative expectations of what is fair and just. In turn, this desire for fairer social media data mining can be located in the context concerns within media and communications studies about the types of media and media practices which can enhance people's efforts to live good lives, as seen, for example, in the work of David Hesmondhalgh (2013, 2014). Hesmondhalgh notes that a core theme in media scholarship is the way that the marketisation of communication and culture inhibits their contribution to living a good life. Given this, how we might work towards 'beneficial experiences of social justice in life as it is currently lived' (2013, p. 8) through engagement with cultural goods becomes a pressing question. To answer it, he turns to philosophy, particularly the work of Amartya Sen and Martha Nussbaum, who seek to identify the conditions under which humans might flourish. These questions, of course, extend beyond media and social media to social life more generally, and are taken up by Andrew Sayer in the book I reference throughout this chapter (Sayer 2011). Here, Sayer argues that ideas about flourishing and well-being are unfamiliar territory for social scientists, and yet they cannot be avoided if we are to attempt to understand how greater social justice might be achieved. These ideas are relevant to the discussion in this chapter: we might argue that social media activity represents an effort to flourish, through engagement with cultural goods, that the mining of social media data is one spectacular form of marketisation and that participants' reflections on the fairness of social media data mining scenarios represents a quest for a good life for social media users. These points in turn raise big philosophical questions, such as what precise notions of fairness are mobilised, what is the range, where do they come from, and what hopes do they reflect? While this chapter has only started the modest process of unveiling this concern about fairness, future research into data mining would benefit from further engagement with concepts like well-being, social justice and fairness, in order to consider whether a better relationship between social media data mining and social life is possible.

WHAT CONCERNS USERS? FAIRNESS, TRANSPARENCY, CONTEXTUAL INTEGRITY

Focus group participants responded in diverse ways to specific examples of social media data mining, echoing the diversity of findings across the quantitative studies discussed above, and their responses appear to be informed by a number of factors—age, nationality, occupation, extent of social media use and prior knowledge of social media data mining all seemed to play a role. The type of data tracked and gathered, the purpose of the monitoring activity, the extent to which data gathered are perceived to be public or private, and views about transparency and informed consent also appeared to inform participants' perspectives. The ways in which users differentiate social media data mining practices and the variation in their perspectives towards them is consistent with the argument of this book that when we talk about social media data mining, we need to differentiate types of data mining, actors engaged in it, institutional and organisational contexts in which data mining takes place, and the range of purposes, intentions and consequences of data mining. Participants' considerations of the examples we put to them are also revealing of the ways in which people evaluate whether, why and how things matter to them (Sayer 2011).

In the focus groups, we found that, at times, participants brought the discussion round to questions of privacy, even though we did not frame our research in these terms. We did not ask them what they thought about privacy in relation to social media data mining, but rather what they thought of social media data mining, and they responded by talking about privacy, among other things. On the one hand, this demonstrates that this issue matters to users. Specifically, a number of participants suggested that social media *feel* private, personal and intimate, even when they are not. This echoes previous research I undertook into one of the earliest forms of online self-representation, the personal homepage (Kennedy 2006). In that research, I found that the 'extraordinarily frank and revealing' (Chandler 1998) content that people included in their homepages resulted from a feeling of anonymity, even though homepages cannot be described as anonymous. Likewise, on social media, users feel that their content is private, because it is personal and intimate, even when, technically, it is not. Data miners' view that public social media data is fair game for mining and analysing, because it is in the public domain, is called into question by what most of these focus group participants said. Given this, data miners (and the rest of us) need to take seriously the fact that social media users sometimes feel that they are operating in private spaces, or sharing information privately, even when they are not. Doing this means putting social media users' views at the centre of debate which aims to advance our understanding of—and our practice in relation to—social media data mining.

But participants' discomfort with the mining of what feels to be or is designated as private data moves beyond an interest in their own individual privacy to a broader concern about whether social media platforms are operating within fair and just norms. Among the variation in participants' responses, a pattern emerged, of focus group participants assessing the fairness of specific practices and arriving at different conclusions about whether practices were fair or not. This common concern about the fairness of data mining practices explains the diversity in responses: differences resulted from distinct evaluations of whether practices are fair. Participants weighed up the attributes of the data in question, thought about informational norms, transmission principles, and rights and responsibilities (of platforms and users), all elements of contextual integrity according to Nissenbaum. But as Sayer (2011) suggests, such language—contextual integrity, lay normativity—is alienating and the lay term 'fairness' is better for capturing what matters to people in relation to social media data mining. Clearly, there is a discrepancy between the practices of the platforms and users' normative expectations, especially with regard to transparency and consent. Participants' consideration of how to ensure greater transparency and support for regulatory measures which would require platforms to communicate simply and clearly about their data mining activities is relevant to those undertaking social media data mining, as acting in ways which users find acceptable is surely important. This raises the issue of how evaluations of fairness might guide data mining practices, and who gets to control what definitions of fairness count. These are tricky, philosophical questions, and I return to them briefly in the book's conclusion.

We might expect that the ordinary kinds of social media data mining that we discussed with our focus group participants and that are the focus of this book do not concern social media users. We might expect that no-one really minds if unknown others know that 'I really like cornflakes', as one of our participants put it. We might also expect that the ubiquitous and everyday mining of social media data normalises data mining, so that people come to expect, and not to question, the mining of their personal, social data. Turow and others (2015b) suggest this happens in relation to retail surveillance and his recent collaboratively authored report, entitled *The Tradeoff Fallacy: how marketers are misrepresenting American consumers and opening them up to exploitation* (Turow et al. 2015a), suggests that something similar occurs in relation to digital data tracking. It is not the case, the authors argue, that people are willing to trade their data for services—this is a 'tradeoff fallacy'—but rather, people feel resigned to data mining, viewing it as a practice over which they have no control and from which it is not really possible to exclude themselves. However, we did not encounter such resignation in our study. On the contrary, we found that some forms, aspects and uses of social media data mining concern users, especially when private data are shared, data are mined without consent, vulnerable subjects have their data mined or are 'acted upon' in other ways, and there is insufficient transparency about the mining of our data. Norms relating to social media data mining may not yet be established, but these things bother users, and as norms become established, it would be good for this to be acknowledged.

Some social media data mining is carried out for what is believed to be the social good, such as the academic examples we included in our scenarios, and the example that one participant gave of his own experience of using data mining in his work of organising community development volunteering. These might be seen as efforts to 'do good with data'. The next chapter focuses more extensively on such efforts, and so brings us back to the questions of whether the technological assemblage of social media data mining can be harnessed for the social good, and what kinds of agency can be enacted in relation to it.

CHAPTER 8

Doing Good with Data: Alternative Practices, Elephants in Rooms

INTRODUCTION

So far, I have focused on a number of sites in which ordinary social media data mining takes place: local, public sector organisations experimenting with insights tools, commercial companies offering social insights to paying clients, and organisations which engage the services of these commercial companies. I have also considered how social media users' views about the mining of their social data might inform thinking about what should concern us about social media data mining. In this chapter, I return to the question of whether social media (and other) data mining can be used in ways that make a positive contribution to social life which I introduced in earlier chapters. I do this by focusing on two fields in which actors might think of themselves, in different ways, as 'doing good with data' (to adapt the strapline of US data visualisation agency Periscopic (http://www.periscopic. com/)).

First I focus on academic social media data mining, as a growing number of social science and humanities researchers seek to use social insights tools, or collaborate with computer scientists and professional data mining practitioners in order to make a positive contribution to knowledge about society. Academic social media data mining is the elephant in the room in the chapter's subtitle, because criticisms of data mining such as those outlined in Chapter 3 are rarely discussed in relation to academic practices. I then turn to the 'alternative practices' of this chapter's subtitle, or forms of data activism, such as open data initiatives, data art and

© The Editor(s) (if applicable) and The Author(s) 2016   189
H. Kennedy, *Post, Mine, Repeat*,
DOI 10.1057/978-1-137-35398-6_8

data visualisation, campaigns for better and fairer legislation in relation to data, and movements which seek to evade dataveillance. While not strictly focused on *social media* data (although some are), activist groups seek to implement data-related arrangements which enable citizens and publics and, as such, they are relevant to this book's focus on possible forms of agency in relation to data mining.

In an article about what they describe as 'interface methods' for digital social research, Marres and Gerlitz (2015) argue that data mining assemblages—including social media data mining processes—are 'complex and unstable'. They draw together 'a diversity of people, things and concepts in the pursuit of particular purposes, aims and objectives' (Harvey et al. 2013) and, as such, they may enable different agencies in different settings. They may serve a range of different agendas beyond big business and big brother, write Marres and Gerlitz, including political, not-for-profit, ethical and research-related agendas. Such ideas about the instability and potential of data mining underlie this book, and they especially inform this chapter's focus on its uses by social researchers and data activists.

I have not carried out empirical studies of data activism and academic data mining along the same lines as research reported in earlier empirical chapters, although, as noted in the introduction, the social media data mining events that researchers and I have attended, including academic events, constitute a kind of empirical data gathering. But it is important to discuss these data mining practices, both academic and activist, for two reasons. First, because the desire to do good with data that underpins the examples discussed here is important in relation to this book's concern with whether social media data mining always inevitably suppresses human well-being, or whether alternatives are possible. Second, they are both consequence and constitutive of the becoming-ordinary of data mining. So in the pages that follow, I provide a brief sketch of both academic and activist fields of data mining practice and I consider them in relation to the questions at the heart of this book. I outline some of the criticisms that have been levelled at academic and activist data initiatives and I argue that, while there are ways in which both can be considered problematic, they are not *only* problematic: they also serve to open up spaces for alternative and better (social media) data mining. We need to recognise both the problems *and* the potential of efforts to do good with data, I suggest, and this requires an openness to considering that critical thinking about data mining might coexist alongside an appreciation of its problem-solving potential.

ELEPHANTS IN ROOMS: ACADEMIC SOCIAL MEDIA DATA MINING

On 29 September 2014, Lev Manovich tweeted that the number of research papers listed on Google Scholar for searches for the keywords 'Facebook dataset' and 'Twitter dataset' were 192,000 and 241,000 respectively. Increasingly, academics in the social sciences and humanities formerly unaccustomed to using quantitative methods are adding social media data mining techniques to their methodological toolsets. As noted in the book's introduction, information about academic research using these methods is usually much more public than is the case with corporate and commercial actors and can be accessed by attending academic conferences, reading books and journal articles and participating in mailing lists and other online discussions. The ethics of such practices are also on the academic agenda and publicly discussed, at least by some. This section provides a brief sketch of academic uses of social media mining, concentrating on the social sciences and humanities, as these fields are most directly concerned with the kinds of social and cultural change that I address in this book. It then considers the same question that has been asked of other forms of social media data mining: should we be concerned about it?

Overview of Academic Social Media Data Mining

It is difficult to identify a precise moment when digital data mining began to be used in social and cultural research. In the European context, one important hub, the Digital Methods Initiative (DMI) at the University of Amsterdam in the Netherlands, dates back to 1999, when it developed tools like the Net Locator (now defunct) and the Issue Crawler for identifying issue-related networks, discussed in Chapter 2. These were not solely for mining social media data, but can be considered precursors to later tools. The DMI now hosts an extensive suite of tools for a range of research into the 'natively digital', as DMI director Richard Rogers (2013) puts it—that is, not only social media, but also blogs, online news, discussion lists and forums, search engines, folksonomies and more. In terms of social media analysis, the DMI has developed tools for doing research on a number of social media platforms, including Wikipedia, Instagram, Tumblr, Twitter and Facebook. Specific examples of tools for extracting data from the latter two major platforms include the Twitter Capture and Analysis Toolset (TCAT) which captures and analyses tweets, and Netvizz, which extracts datasets from Facebook.

But the DMI is not simply a developer of software for researching the natively digital. Like other groups of social science and humanities researchers playing a leading role in the development of digital methods, it also engages in the critical interrogation of whether the tools of the internet can be repurposed for critical social and cultural research. In order to do this, and to deconstruct the political and epistemological entanglements of digital tools, Rogers suggests that it is necessary to:

> Follow the methods of the medium as they evolve, learn from how the dominant devices treat natively digital objects, and think along with those object treatments and devices so as to recombine or build on top of them. (Rogers 2013, p. 5)

By following the methods of the medium, Rogers suggests, new tools can be developed which intervene critically in digital data assemblages.

A growing number of hubs and centres deploy digital methods to mine social media and other online data for social research, including the médialab at Sciences Po in Paris, founded in 2009 with the aim of connecting the social sciences and humanities with digital tools. Like the DMI, it works to develop tools for scraping, mining and analysing various online sources, and it currently hosts the visualisation software Gephi, discussed in Chapter 2. In Australia, the Digital Media Research Centre at Queensland University of Technology has established itself as a significant player in the field of digital methods. The social media mining landscape in the North American academy is harder to map from across the pond, but prominent centres include: the Software Studies Initiative at the City University of New York (CUNY) led by Lev Manovich; SoMe, the Social Media Lab at the University of Washington; the Annenberg Innovation Lab at the University of Southern California; and the Social Media Research Lab at Ryerson University in Toronto. In some cases, former academics have made the leap into commercial data mining, such as Stu Shulman who set up Texifter, developer of DiscoverText which is increasingly widely used by academics, and Mark Smith, now of the Social Media Research Foundation which developed NodeXL, also discussed in Chapter 2. Students and collaborators from the centres mentioned here are taking digital methods and related concerns into the wider academic community and other digital methods hubs are beginning to emerge, such as the Centre for the Study of Innovation and Social Process at Goldsmiths,

University of London and the Centre for Interdisciplinary Methodologies at the University of Warwick. Also in the UK, computer scientist Mike Thelwall at the University of Wolverhampton has been developing free social media research tools for social scientists for some time (http://www.scit.wlv.ac.uk/~cm1993/mycv.html), and at the time of writing, he is collaborating on an innovative project to develop a tool for the analysis of social media images which will be freely available for academic research (http://visualsocialmedialab.blogspot.co.uk/).

As social media data mining methods become more widely used by academic researchers, studies deploying these methods are also multiplying, many of them carried out by the same people developing and thinking about the tools discussed above. Digitally inclined social and cultural researchers who consider the acquisition of these kinds of skills beyond them are increasingly engaging in collaborations with computer scientists. The Web Science Institute (http://www.southampton.ac.uk/wsi/index.page) at the University of Southampton is a good example. It brings together multidisciplinary expertise from across the computer and social sciences to explore socio-digital issues. Such endeavours are not entirely new, of course. But while the application of computational expertise to social and cultural inquiry pre-exists our current datafied times, as data mining becomes ordinary, so the scale of such practices grows. To accompany these developments, how-to books aimed at social scientists and humanities scholars have begun to emerge. These include: Attewell and Monaghan's (2015) *Data Mining for the Social Sciences: an introduction*; Russell's (2013) *Mining the Social Web: data mining Facebook, Twitter, LinkedIn, Google+, GitHub and more*; Vis and Thelwall's *Researching Social Media* (2016/forthcoming) and Zafarani et al.'s (2014) *Social Media Mining: an introduction* (see also Kawash 2015; Szabo and Boykin 2015).

Concerns and Criticisms

Given the proliferation of academic data mining tools, techniques, centres, projects and textbooks, should we be concerned about academic social media data mining? Although not as high profile as more spectacular forms, academic social media research has come onto the public radar and raised some concern. An article in *USA Today* entitled 'Social media research raises privacy and ethics issues' (Jayson 2014) suggests that social media users may not like the fact that their behaviour on social media platforms is under academic scrutiny. The article cites a paper by social psychologist

Ilka Gleibs about the use of social media platforms as sites for field research, in which she warns that, in the context of academic social media analytics, 'Facebook is transformed from a public space to a behavioural laboratory' (Gleibs 2014, p. 358). It goes on to quote Gleibs' warning to social media users that they should 'be aware it is a space that is watched'. Also alert to the fact of growing academic research on social media platforms, the platform from which much social data is extracted, Twitter, warns its users of the possibility of academic researchers mining and analysing Twitter data:

> Your public user profile information and public Tweets are immediately delivered via SMS and our APIs to our partners and other third parties, including search engines, developers, and publishers that integrate Twitter content into their services, and institutions such as universities and public health agencies that analyse the information for trends and insights. When you share information or content like photos, videos, and links via the Services, you should think carefully about what you are making public. (Twitter 2014)

In Gleib's article, she argues that research based on mined social media data is not exempt from the usual ethical standards which apply to all academic research with human subjects, such as the requirement that informed consent is obtained from participants. We might expect infrastructural arrangements within universities, such as ethical review committees in the UK and Institutional Review Boards (IRBs) in the USA to ensure that such standards are maintained with regard to social media research. These administrative processes form part of the broader scholarly endeavour to ensure that codes of research practice committed to the excellence, honesty, integrity and accountability of research and researchers are followed. As the 2012 version of the Association of Internet Researchers (AOIR)'s Ethical Guidelines puts it, 'the basic tenets shared by these policies include the fundamental rights of human dignity, autonomy, protection, safety, maximization of benefits and minimization of harms, or, in the most recent accepted phrasing, respect for persons, justice, and beneficence' (that is, protection of the welfare of research participants) (AOIR 2012, p. 4). Ethical guidelines generally recommend that in academic research with human subjects, benefits should outweigh risks, all parties should be protected from harm, especially vulnerable subjects, participants should give informed consent and have the right to withdraw, and what they share should be treated confidentially. But as the AOIR's guidelines acknowledge, internet-based research methods like social media data mining throw up all sorts of questions that offline research has not needed to address, such as: is it right to characterise social media data mining as research with human subjects? How do we define human subjects in a big data age? Is informed consent necessary and possible when accessing large-scale datasets?

Some university ethics committees seek direction from the AOIR's Ethical Guidelines to try to address the particular ethical issues that arise in internet research. These guidelines acknowledge the dynamic character of internet research and therefore offer a set of principles to be interpreted on an inductive, case-by-case basis, recognising the complex array of factors that come together in ethical decision-making, including 'one's fundamental world view (ontology, epistemology, values, etc), one's academic and political environment (purposes), one's defining disciplinary assumptions, and one's methodological stances' (AOIR 2012, p. 3). In taking this position, the AOIR may seem to miss an opportunity to provide strong ethical direction to internet research. However, despite this caution with regard to the ethical issues that internet research raises, the AOIR guidelines make some robust assertions. First, researching digital data is researching with human subjects, 'even if it is not immediately apparent how and where persons are involved in the research data' (2012, p. 4). Second, the historical public/private distinction does not hold in digital spaces, so it is important to consider what 'ethical expectations users attach to the venue in which they are interacting, particularly around issues of privacy' (2012, p. 8). Third and relatedly, online data, including social media data, cannot be assumed to be intended for use as research data, so researchers need to ask themselves 'what possible risk or harm might result from reuse and publication of this information?' (2012, p. 9).

There is significant overlap between these assertions in the AOIR ethical guidance and Nissenbaum's (2009) argument that contextual integrity is needed with regard to digital data tracking, discussed in previous chapters. For Nissenbaum, contextual integrity means being attentive to the expectations that people have in relation to what will happen to their data and ensuring that data flow appropriately in data mining practices. So the issues that are of concern to users (who mines data, in what contexts and for what purposes), and which were not widely acknowledged by the people whose data mining formed the focus of previous chapters, are also relevant to academic data miners. These are ethical issues from which academic researchers are not exempt.

Researchers using social media data mining methods, who have subjected their research plans to the scrutiny of ethics committees and have been given the green light to proceed, may well think that their research is, ethically speaking, beyond concern. The AOIR guidelines are thorough, but not all university ethics committees or IRBs know they exist, never mind adhere to them, meaning some projects are likely to slip through the ethical net without consideration of the full range of issues that the AOIR guidelines raise. As one communications scholar using social media mining techniques said to me in a conversation, 'I hold myself to much more rigorous ethical scrutiny than IRBs do.' But, more significantly, ethical review procedures do not attend to the full range of concerns that social media data mining raises—they do not exist in order to do this. Whether academic research with social media mining leaves the problems that datafication brings (discussed throughout this book) unchallenged is a concern that lies far beyond the remit of ethical review. Such procedures do not subject academic social data mining to the same critical scrutiny to which other forms have been subjected because this is not their purpose. Another strategy is needed, one which interrogates academic practices critically.

Van Dijck's 2014 article in the special issue of *Surveillance & Society* on big data surveillance attempts to do this, to interrogate academic social media data mining critically. Here van Dijck argues that the logic of datafication has found a place 'in the comfort zone of most people' (2014, p. 198). As a result, governments, businesses and academics participate in dataveillance practices which exploit the becoming-ordinary of datafication and of social media logic. 'Datafication as a legitimate means to *access*, *understand* and *monitor* people's behavior is becoming a leading principle, not just amongst techno-adepts, but also amongst scholars who see datafication as a revolutionary research opportunity to investigate human conduct' (2014, p. 198), she writes. Thus van Dijck argues that it is not just business and government that hail social media data as 'the holy grail of behavioural knowledge', but academic researchers too. In the rhetoric of all of these groups, data is considered to be equal to people—but not in the encouraging way that the AOIR Ethics Guidelines acknowledge the intimate relationship between people and their data and the expectations of appropriate data flows that accompany this relationship. Rather, data are considered to offer direct access to knowledge about the social, van Dijck claims, and platforms are seen as neutral facilitators of this knowledge: there is a widespread lack of recognition of the work that platforms do in shaping the action that takes place within them (what could be called 'platform effects', following Marres' (2015) notion of 'device effects'). Van Dijck argues that 'information scientists often uncritically adopt the assumptions and ideological viewpoints put forward by SNSs and data firms' (2014, p. 201) and in doing so, are complicit in 'maintaining credibility of the ecosystem as a whole.' This threatens researchers' role in building social trust, because traditionally 'a paradigm resting on the pillars of academic institutions often forms an arbiter of what counts as fact or opinion, as fact or projection'. To counter these problems, van Dijck argues:

> The unbridled enthusiasm of many researchers for datafication as a neutral paradigm, reflecting a belief in an objective quantified understanding of the social, ought to be scrutinized more rigorously. Uncritical acceptance of datafication's underpinning ideological and commercial premises may well undermine the integrity of academic research in the long run. To keep and maintain trust, Big Data researchers need to identify the partial perspectives from which data are analyzed; rather than maintaining claims to neutrality, they ought to account for the context in which data sets are generated and pair off quantitative methodologies with qualitative questions. (2014, p. 206)

It is important to clarify which researchers van Dijck is writing about here. She uses the phrase 'information scientists' to delineate them, and papers cited come from journals and conference proceedings in this field, including the IEEE International Conference on Social Computing and the CHI *Conference (on Human Factors in Computing Systems) Proceedings*. In the absence of any systematic mapping, it is hard to know how widespread the ideologies that van Dijck notes are among information scientists or within other disciplines. Elsewhere, and as noted in an earlier chapter, Boellstorff (2013) has argued that those working with big data understand their datasets as limited representations of the world, conditioned by the theories that frame them and Havalais (2013) has proposed that many scientists accept that data are influenced by the processes by which they are collected. These assertions seem to suggest some recognition of some of the issues that van Dijck highlights, at least among data scientists.

Do the criticisms that van Dijck poses apply to social scientists and humanities scholars? The question is difficult to answer. On the one hand, as seen in the quote below, taken from the introduction to an edited collection of studies of Twitter and society and written by a group of social scientists, the view that content found on Twitter can be understood simplistically as 'research data' has some traction in this field:

> The substantial amount of content generated and shared by Twitter users, from individuals to institutions, also opens up exciting new research possibilities across a variety of disciplines, including media and communication studies, linguistics, sociology, psychology, political science, information and computer science, education, and economics. There remains a significant need for the further development of innovative methods and approaches which are able to deal with such new sources of research data, and for the training of a new generation of scholars who are deeply familiar with such methodological frameworks. (Weller et al. 2013, p. xxxi)

On the other hand, and as seen on the next page of the chapter cited above, among social and cultural researchers, there is some acknowledgement of the need to unpack these assumptions and to address what van Dijck calls 'the fundamental epistemological and ontological questions' (2014, p. 206) that these methods bring with them. Here Weller et al. suggest that the onset of computational social science requires 'a significant amount of further thought' into 'the conceptual, methodological, and ethical frameworks which we apply to such work' (2013, p. xxxii). In the remainder of this section, I discuss two examples of social media research which attempt to do this. There are many more examples, but I select these two because they are illustrative of good, critical, reflective practice and of two tendencies. I characterise the first as drawing attention to how data are made and shaped, and the second as exploring the underdeterminacy of data mining apparatuses described by Marres and Gerlitz (2015). Both, I suggest, represent efforts to un-black-box social media data mining.

Un-Black-Boxing Social Media Data Mining

In a paper entitled 'Working Within a Black Box: transparency in the collection and production of big twitter data', social media researchers Kevin Driscoll and Shaun Walker (2014) set out to make visible the different actors in the data making process. The authors argue for a systematic approach to describing clearly the conditions under which data are made—in this case, Twitter data. They insist that, in order for readers to make sense of such data, information is needed about how data are collected, stored, cleaned and analysed. What's more, the conditions of data making need to be evaluated. This approach, they argue, would constitute the establishment of shared standards for reporting research which has used social media mining methods. They use as evidence an experiment they undertook in obtaining data about the same topic from two different sources: Twitter's publicly accessible streaming API and its 'fire hose' provided by the Gnip PowerTrack, a commercial service from one of Twitter's corporate partners, which is costly in terms of both licence fee and required supporting infrastructure. Not surprisingly, the two different approaches found different data, despite identical searches. Through this example, Driscoll and Walker unveil the instability of social media data, the infrastructure required to access them and the expertise required to make sense of them. Thus they show how the conclusions that researchers draw from such studies are mediated in many ways, including 'by the contours of our local data management systems' (2014, 1761).

Driscoll and Walker's article attempts to demonstrate the ways in which social media platforms are not neutral facilitators of access to natural traces of social phenomena. They make the opacity of social media platforms the focus of their research, not something to be obscured from view in the reporting of their 'findings'. They are motivated, like Bruns and Burgess (2011a, b, 2012), who they cite, to develop a set of standards for reporting on research with social media data and so to work against the black-boxing of social media platforms like Twitter. They are not the only writers to undertake such an endeavour, and other, similar articles can be found. One example is Farida Vis's contribution to the 2013 special issue of the online journal *First Monday* on big data (and beyond). In her article, Vis makes visible the interweaving of programme APIs, researchers and tools in the production of data. I highlight the paper by Driscoll and Walker not because it is unique, but because it provides a good example of this move to un-black-box social media data mining processes.

The second example of un-black-bosing involves embracing and experimenting with the instability of data mining. Researchers working with and within the DMI suggest that academics using social media data mining methods should accept the volatility and uncertainty of social media data and work with it. For example, in a paper entitled 'Interface methods: renegotiating relations between digital social research, STS and sociology', Marres and Gerlitz (2015) ask and answer these questions about the opportunities opened up for critical and creative methods development by the proliferation of digital data mining tools:

Should it be our aim to clear up the methodological ambiguities opened up by digital analytics, and differentiate between the journalistic, commercial, everyday, governmental use and the sociological implementation of these tools? Or is there something productive about these very resonances and suggested affinities? We will propose that there are decisive advantages to affirming the ambivalence of digital analytics according to which data tools are both similar and different from sociological research techniques. (2015, p. 4)

Marres and Gerlitz propose that asking whether particular social media data mining tools and platforms are ethical or fit for purpose is not productive, and that instead social researchers should embrace their underdeterminacy. These are, according to Marres, 'multifarious instruments' (Marres 2012) which serve multiple purposes and therefore should be treated as objects with which to experiment. 'Instead of asking what the capacities of social digital methods are, and deciding with which agendas they are and are not in alignment, we advocate experimental inquiry in to what makes their deployment productive for social inquiry', write Marres and Gerlitz (2015, p. 3). In other words, these authors propose that we need to shift our attention from what methods are and what their limitations are to how they can 'become intellectually relevant through specific deployments' (2015, p. 12). After all, if a tool can serve multiple purposes, it cannot be defined narrowly, they suggest. The argument that Marres and Gerlitz develop, here in this article and elsewhere with other collaborators (for example Gerlitz and Rieder 2013), would seem to be an argument against the main purpose of this book. We should not be concerned with what should concern us about digital data methods, they seem to suggest, and we should instead experiment with their use. But I think that both are needed. Given their effects—traced throughout this book—as well as their instability, we need to interrogate data mining methods normatively and experiment with their potential to do good.

Marres, Gerlitz and their collaborators have carried out a number of social research experiments, some of which demonstrate how social media data mining might be mobilised to critique social media data mining itself. In a keynote talk at the DMI Winter School in 2015, entitled 'A critique of social media metrics: the production, circulation and performativity of social media metrics', Gerlitz argued that digital methods can be understood *not only* as repurposing the methods of the media that they investigate, as Rogers (2013) proposes, but also as tools with which to critically enquire into their own practice (Gerlitz 2015). As an example, she referred to her study with Bernhard Rieder (Gerlitz and Rieder 2013) of what she describes as 'intervening variables' in hashtag use on Twitter, or factors that affect the usage of particular hashtag types. They found that specific devices cater for specific hashtag uses: calls to action, for example, or promotional hashtags, are more common on some devices than others. In this context, devices become intervening variables—there are device effects, as well as the platform effects discussed elsewhere. Gerlitz and Rieder conclude that hashtags are not all created equal: devices come with different use practices and imply different regimes of being on Twitter, and these in turn influence hashtag use. Examining the variables that intervene in hashtag use with social media methods enables this kind of critique, suggested Gerlitz in her keynote. Just as metrics participate in that which they seek to measure, so too digital methods participate in the production of metrics. But, as this example shows, argued Gerlitz, such methods also hold the potential for 'critical numbering practice', because they can unveil the very platform and device effects which shape and make data. Here, it is not social media data that is studied, but social media data methods.

The ethics of data mining that have surfaced throughout this book come into sharp focus in academic settings, in which research is expected to be undertaken in ethically rigorous conditions. But despite the work of bodies like the AOIR to bring internet research ethics to prominence, ethical considerations remain somewhat in the shadows, as they do in the other contexts discussed in previous chapters. In Chapter 3, I noted that dominant concerns about spectacular forms of data mining might not apply to more ordinary manifestations, but we have seen in discussions with users and with data mining practitioners some unease about the mining of social media data. Whether we call this a privacy issue, or an issue relating to the intimate and personal character of social data, is not so relevant. What matters is that users seem to expect contextual integrity and that data mining generally ignores these expectations. The perception that social media data in the public domain is open to use needs to be challenged, as the AOIR guidelines suggest, but the interpretative flexibility of social media data, data analytics and data mining ethics makes it difficult for practitioners to decide what is right and how to act (and so does the desire for numbers, I have argued in previous chapters). Marres and Gerlitz suggest that we should postpone our judgement about what is good and ethical in social media research, because of this interpretative flexibility, this instability. These methods are too new, they argue, for us to make normative decisions about them, and experimentation with what they enable is needed before we arrive at any categorical conclusions. Their own experiments show that data mining can be undertaken in ways which reflect critically on data mining itself and so contribute to knowledge about its operations. But I agree with only half of their argument: we should experiment with what social media data mining makes possible and whether we can do good with it, but, because data mining is so integral to the logic of datafication and to data power, and because these are having effects now, we need to do the difficult job of judging, normatively and critically, now, not in the future. I do this in the next section, in relation to other efforts to repurpose data mining and do good with data: data activism.

ALTERNATIVE PRACTICES: DATA ACTIVISM

(I)t is crucial that access to the underlying data remains open and free, so that actors that do not have the economic means to pay for such data, such as activist groups, consumer cooperatives or other non-profit organisations will still be able to operate and construct devices. (Arvidsson, 2011, p. 22)

In the above quote from an article about the convergence of value and affect in the information economy, Adam Arvidsson (2011) proposes that 'the wrong people' lose out if digital data are not public. With these words Arvidsson captures the interests of many data activists, who are the focus of this section (although not the focus of Arvidsson's paper). Below, I frame examples of data activism, including open data initiatives, re-active data activism, citizen science, data art and visualisation, as efforts to do good with data. I highlight some of the ways in which data activism has been criticised and argue that it is unhelpful to understand data activism only in these critical terms. I start with the open data movement.

The Open Data Movement

Open data initiatives have gained significant traction in recent years. Governments around the world have launched hundreds of portals to open up their data, and major multinational corporations and philanthropic organisations have spent large sums of money to support its growth. Open data initiatives, movements dedicated to opening up government data or to free and open source software (sometimes called FLOSS, with L for *libre* added in) can all be understood as efforts to reverse the commercialisation of data and information that occurred in previous years (Bates 2013). Open Government Data, for example, is an information policy proposal allowing the access and re-use of datasets produced by public institutions—these might relate to meteorology, land use, public transport, company registration or government spending (but not social media data). Under open data principles, these datasets should be accessible and usable by everyone, 'free of charge, and without discrimination' (Bates 2013).

There are hundreds of open data initiatives across the globe, from national and city-wide open data groups, to networking organisations like the Open Data Institute (http://opendatainstitute.org/) and the Open Knowledge Foundation (https://okfn.org/) in the UK. One widely cited example is Transparent Chennai (http://www.transparentchennai. com/) in India, which mines and brings together data about civic issues faced by marginalised communities in the city of Chennai. In doing so, Transparent Chennai aims to address what it sees as the problem of a lack of data which allows the city government to evade its responsibilities. The group creates data maps to communicate civic data to the city's residents and so enable them to have a voice in city planning and governance. For example one project focuses on sanitation and public toilets (http://www.transparentchennai.com/public-toilets-and-sanitation/). Noting the importance of public toilets for slum-dwellers, the homeless and people working in the informal sector, and yet their poor supply in the city, the group collected data on their location and governance in order to inform public debate and develop public understanding of who is accountable for the city's sanitation.

Open data has taken off to such an extent that technology writer Alex Howard argues that local governments are becoming data suppliers, and there is a growing expectancy of transparency when it comes to government data (Howard 2011). However, the Open Data Barometer (http://barometer.opendataresearch.org/report/analysis/rankings. html), a part of the World Wide Web Foundation's efforts to uncover the prevalence and impact of open data initiatives around the world, states in a report in January 2015 that there is a lack of open and accessible data on the performance of key public services. The report concludes that there is a pressing need for further investment in open data capacity-building, training and support.

Some civil society actors see the opening up of public datasets as the democratisation of data, according to Bates (2013), as it allows the kind of access that boyd and Crawford argue is worryingly absent in relation to much big data. Taking a balanced perspective, Baack (2015) highlights how open data movements represent an intriguing coming together of the two contradictory tendencies that are at the heart of this book—that is, the problems and potential of data mining. On the one hand, he notes, open data movements depend on datafication for their existence, and all the troubling consequences that this phenomenon brings with it. On the other hand, they also depend on the democratic practices and values of open source culture, including advocacy of transparent and collaborative forms of governance and the right to access and distribute knowledge. Baack explores what this unusual convergence reveals about the relationship between data and agency, with the former tendency, datafication, arguably suppressing the possibility of public agency in relation to data and the latter tendency, open source practices, arguably enabling it. He concludes that datafication supports, rather than undermines, the agency of data activists.

The incongruous forces which come together in open data movements can be traced in the six 'visions' of open data that Bounegru and others have traced in media discourses on the topic. According to Bounegru et al. (2015), media discourses construct open data as relating to:

1. transparency, anti-corruption and accountability
2. democracy, participation and empowerment
3. public service delivery, decision-making and policy-making
4. efficiency and waste
5. unlocking innovation and enabling new applications and services
6. economic growth and new business.

The contradictory discourses of commercial benefit and public empowerment evident in these visions suggest that opening data up does not lead only to their democratisation, as a number of writers have noted (for example Bates 2013; Rae 2014). As urban data analyst Alisdair Rae (2014) points out in a blogpost, we should not assume that open data are a public good per se; rather, we need to ask three questions in relation to specific open data projects. First, we need to ask 'opened by whom?' The answer to this question will tell us something about the intentions of those opening up data. Second, we should ask 'opened for whom?' as the answer to this question enables assessment of whether the data in question can be considered 'good' open data in the sense outlined by the Open Data Institute—that is, capable of being linked to, easily shared and talked about and available in a standard, structured format so they can be easily processed (this question also draws attention to the need for data intermediaries, as not everyone has the expertise to make sense of raw data). Finally, to be sure that open data is not simply a solution in search of a problem, it is necessary to ask 'open for what?', as the absence of open data standards can mean that not much can be done with data made 'open' in only limited ways. I return to these questions—and some answers to them—below, but first, I outline some other forms of data activism.

Re-active and Pro-active Data Activism

Stefania Milan uses the categories 're-active' and 'pro-active' to classify types of data activism. Re-active data activism, she suggests, involves individuals resisting the collection of personal data, and pro-active data activism relates to social movements' use of big data to foster social change (Milan 2014). According to this definition, open data movements might be considered as examples of pro-active data activism. Re-active data activism includes the use of strategies which aim to challenge or resist the mining of one's own online data, such as abstention—that is, not participating in online platforms which require users to share personal data. Another example is setting up false accounts, or producing false or misleading data which either confuses data miners or requires them to invest time in separating bad data from good. Brunton and Nissenbaum (2011) call this latter tactic 'obfuscation', which they see as a form of vernacular resistance. Another re-active strategy is to employ one-time use tools or use services like justdelete.me, a directory of direct links to delete accounts from websites and social media platforms. Other approaches involve using tools which track and block tracking itself, like Ghostery (https://www.ghostery.com/en/), a browser extension for Firefox which alerts users to trackers and web bugs, Collusion (http://collusion.toolness.org/) which produces real-time visualisations of the entities that track users across the web, or Disconnect (https://disconnect.me/), which visualises and blocks sites which track users' search and browsing history. Other 're-active' tools allow users to take control of their data in different ways. Tor is a piece of software that helps users defend against data monitoring via onion routing (re-directing online activity via a complex network and therefore making it harder to track). PGP (Pretty Good Privacy) encrypts data, making it hard to decipher. Or users can select alternative platforms which claim not to track or share data, like Riseup.net, a non-profit internet and email service provider (although for non-profits like this, their user data might be tracked and mined without their knowledge). Another example is the recently launched Ello, an alternative social network (to Facebook), which claims it will never sell data to advertisers or show ads. Duck Duck Go is an alternative search engine which promises to protect privacy. Individual users' court actions against what they see as breaches of their data rights by social media and other platforms also fit into this category.

There are some limitations to Milan's categorisation. First, it separates practices into two categories which are in reality blurred and overlapping—using some of the technologies listed here might be re-active, but developing them is pro-active, something which is not accounted for in this classification system. Second, her definition of pro-active data activism prioritises social movements with non-data-related agendas mobilising data for social change, so that activism advocating for change relating to *data themselves* is not included. Yet, as we will see below, there are multiple forms of *data-focused* data activism. A third problem is that Milan suggests that an important distinction between data activism and other, offline forms of activism is that the former involves ordinary people whereas the latter does not. I question this assertion below, and refer to other writers who have done the same. Nonetheless, Milan's categories provide some assistance in mapping data activism, and so I use them here, discussing some more examples of pro-active data activism below.

Other kinds of pro-active data activism around open data, in addition to those discussed above, include projects like Open Corporates (https://opencorporates.com/), which aims to bring together and so make public data relating to commercial companies which operate across the globe, so that such data are open and accessible, not hidden from public view. It was built by developers involved in other efforts to open up data, including TheyWorkForYou (in New Zealand), WhosLobbying.com, OpenlyLocal.com and OpenCharities.org and, at the time of writing, it contains data relating to 55 million companies in 75 jurisdictions.[1] Other examples adopting data activism models include

---

[1] Thanks to Dan McQuillan for pointing out some of these to me.

Hack4YourRights initiatives, gatherings which create apps, visualisations and other technologies to make transparent government surveillance activities. The EU's Hack4Participation explores how to get EU citizens more involved in EU policy-making and enable the better analysis of policy-making processes. It sometimes results in the development of data-related policy: policy relating to Net Neutrality was developed at a German policy hackathon and the Icelandic Modern Media Initiative operates in this way, adopting a strategy not of lobbying, but of writing media law which is relevant to the current digital age (for example in relation to privacy) (Hintz 2014). These events not only focus on data-related issues but also adopt some of the mechanisms of data activism, such as: DIY self-production; using open platforms to crowdsource expertise (legal and technical); aiming to open up the policy-making process and enhance participation in it.

Citizen Science—crowdsourcing the collection and analysis of data, the development of technologies, and the testing of phenomena—also operates within the model of opening up data to broad publics. ExciteS (http://www.ucl.ac.uk/excites) (Extreme Citizen Science) at University College London in the UK describes itself as a 'situated, bottom-up practice that takes into account local needs, practices and culture and works with broad networks of people to design and build new devices and knowledge creation processes that can transform the world'. Other initiatives adopting the 'do good with data' spirit of data activism include those that match pro bono data scientists with charities and non-government organisations (NGOs) in need of expertise on how to collect, analyse and visualise data to help them achieve their goals. One example is DataKind (http://www.datakind.org/), whose website makes the bold claim 'we harness the power of data science in the service of humanity'. Launched in 2011 with headquarters in New York City and branches elsewhere in the US, the UK, Ireland, Bangalore and Singapore, DataKind puts data experts together with social groups to address humanitarian problems. Completed projects share data about homelessness, child poverty and international human rights case law. Similar initiatives include 'Code For' events (Code For Europe (http://codeforeurope.net/), Code For Africa (http://www.codeforafrica.org/) and so on), a series of one-off, agile meetings in which teams attempt to solve 'local civic challenges' in response to local needs. As in the other examples discussed here, these events are motivated by a desire to enact citizen-driven openings up of data.

Most of these initiatives do not focus on social media data. Indeed, open data movements often subscribe to the belief that personal data and data that are deemed to be security-related should be excluded from calls for datasets to be made accessible and available (Baack 2015). But one project which has taken on the challenge of thinking about how the strategies discussed here might be mobilised in relation to social data is 'Our Data, Ourselves', led by Mark Coté and Tobias Blanke at Kings College in London (Pybus et al. 2015). This project confronted questions of individual agency in relation to social data, given what the researchers saw as asymmetrical power relationships with regard to who gets to own the social data that we are all active in producing. The project team explored how gaining access to one's own social data might augment agency, working with young coders to create apps to intervene in social data produced on mobile phones. Examples of apps created by the young coders include one designed to highlight the frequency of data tracking through audio alerts and another which produced graphs demonstrating the relationship between social media platform usage and the frequency of data mining. Blanke et al. articulate their motivations for undertaking this project as a desire to create a big social data commons, or a framework for data sharing available for wide community usage. To achieve these aims, they used the Open Knowledge Foundation's CKAN platform, which they describe as an 'alternative database' 'dedicated to greater access and openness', 'respectful of privacy concerns' (though it is not clear how such respect is built in to this technical system).

A different type of data activist from those discussed thus far is the trickster, exemplified in the Anonymous movement, a loose network of (h) activists, identifiable by the Guy Fawkes masks worn in public. Anonymous coordinates protests and hacks against anti-digital piracy campaigns and government agencies and corporations, among others. Broadly, they oppose internet censorship and control, including the control and surveillance of data. Data are integral to their activism in these ways, and also in their use of data as a weapon—one of their strategies is to release (sometimes embarrassing) private data about the people and groups who are their targets. The title of Coleman's book on the movement alerts us to some of its approaches: *Hacker, Hoaxer, Whistleblower, Spy: the many faces of Anonymous* (2014). Evading dataveillance is key to their operations, and in this way too they are data activists.

Other affiliations of tricksters include hacker group LulzSec and the antisec coalition, a movement opposed to the computer security industry, which Dan McQuillan defines as counter-movements to new forms of exclusionary power in times of datafication. Drawing on Agamben, McQuillan (2015) argues that the ubiquity of data mining and its consequences produce a perpetual 'state of exception': data mining operations increasingly operate with coercive force on populations, exempt from substantive legal controls. Anonymous, LulzSec and antisec provide one answer to the question of how we might counter these dominant forces, as they seek to disrupt apparatuses of control without themselves engaging in the 'cycle of law-making and law-preserving' (2015, p. 571). Similarly, McKelvey et al. (2015) explore possibilities for resistance under circumstances in which, they argue, traditional 'logical' forms of antagonistic communication are not the answer. Focusing on the work of internet activist groups 4Chan, an image-based bulletin board sometimes linked to internet hijack pranks, and the Deterritorial Support Group (DSG), an anti-authoritarian internet-based political grouping, McKelvey et al. propose that in permanent states of exception which operate outside the law, one form of 'action from the outside' is to make no sense, 'since logical forms of antagonistic and subversive communication are merely inputs that feed both bots making stock market decisions and governments devising new modes of digital control' (2015, p. 586). They give this example of making no sense from DSG:

> When asked by liberals 'Do you condone or condemn the violence of the (often private property destroying and occasionally violent) Black Bloc?' we can only reply in unison 'This cat is pushing a watermelon out of a lake. Your premise is invalid.' (Nesbitt, 2012: np, cited in McKelvey et al. 2015, p. 587)

Or, as Michael Serres puts it in *The Parasite* (2007), by being pests, these minor groups might become important players in public dialogue about data, surveillance and control.

Artistic projects which aim to 'do good with data' also seek to intervene in relationships of data power. Examples include the work of Adam Harvey and Undisclosed, a New York-based, research-led art studio which use artistic strategies to shine a spotlight on issues relating to privacy and surveillance. One illustration of their interventions is the Privacy Gift Shop (http://privacygiftshop.com/), a pop-up store selling things like OFF Pocket, a privacy accessory for mobile phones which blocks wireless signals, and Stealth Wear, clothing that shields against thermal imaging, a surveillance technology widely used by the military. Another project of theirs is CV Dazzle, which uses 'avant-garde' hairstyling and makeup to shield against face-detection and facial recognition technology. Benjamin Grosser's Facebook Demetricator (http://bengrosser.com/projects/facebook-demetricator/) also falls into this category of data activism through artistic practice. Concerned about the metrification of social and cultural life, as discussed in earlier chapters (Grosser, 2014), Grosser developed the Demetricator to remove statistics from Facebook and so open up the possibility of experiencing the platform free from the tyranny of numbers.[2]

Some data visualisers also aim to 'do good with data'. As acknowledged above, this phrase, used in the chapter's title, is the strapline of visualisation agency Periscopic (http://www.periscopic.com/). Periscopic are best known for making a widely circulated animated visualisation about years lost to gun-related death in the US (http://guns.periscopic.com/?year=2013), from which a screenshot is shown in Figure 8.1. In this visualisation, Periscopic mobilise data to intervene aesthetically in debates about gun laws. For Periscopic and other visualisation designers, the visualisation of data is seen as one means by which to promote data transparency and awareness and so do good with data. This is in part because the main way that people get access to data is through visualisations: 'data are mobilized graphically', say Gitelman and Jackson (2013, p. 12). This idea that visualisation can promote data awareness and transparency can be traced back to the work of Otto and Marie Neurath in the mid-nineteenth century and their development of a graphical language called Isotype, a visual way of representing quantitative information via icons (Zambrano and Engelhardt 2008). The Neuraths believed that 'visual education is related to the extension of intellectual democracy within single communities and within mankind' (Neurath 1973, p. 247) and put their ideas about the power of visualisation into practice in museums they directed, such as the Museum of Society and Economy in Vienna. Belief in visualisation's capacities is brought up to date in contemporary projects like the Roslings' GapMinder (http://www.gapminder.org/world), which describes itself as 'a modern "museum" on the Internet' aiming to promote global sustainable development by visualising related data.

---

[2] See also DMI collaborations with artists, leakygarden.net and elfriendo.com, experiments in composing publics through preferences rather than through demographics, and hence described by Rogers (2013) as 'postdemographic'.

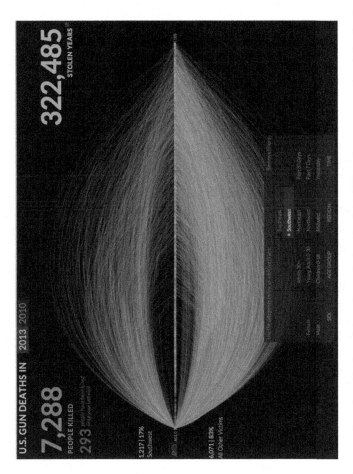

Figure 8.1   A screenshot from Periscopic's animated visualisation of lives lost to gun-related deaths (source: http://guns.periscopic.com/?year=2013, accessed 20th December 2012)

The efforts of other contemporary visualisers can be seen in this way too, such as Stefanie Posavec's 'Open Data Playground' (http://www.stefanieposavec.co.uk/data/#/open-data-playground/), floor-based games which give people the opportunity to physically 'play' with open datasets and so make sense of the data as they see fit.

Doing Good, Or Doing Bad, Through Data Activism?

None of the examples of data activism discussed above are without problems, nor do they represent simple, straightforward and successful subversions of data power. Some of the re-active forms of data activism make it possible to evade data mining on an individual level, but these acts of agency do not confront structures of data power; these are left intact. Criticism has also been levelled at pro-active forms of data activism. For example, writing about open data, Bates (2013) locates some of the problems that she perceives with the rhetoric and practice of related initiatives within a broad political context, building on Braman's argument that information policy—including that which relates to data—is a tool of power in informational states. Drawing on interviews with key policy-makers and activists in the open data movement in the UK, Bates argues that in the case of open government data, Braman's (2006, p. 1) claim that 'governments deliberately, explicitly, and consistently control information creation, processing, flows, and use to exercise power' holds true. In this sense, Bates' research confirms the argument made by Gurstein (2011) and others that open data empowers the already empowered, and as such, does not fulfil its democratising promise. Open data projects, hackathons and (h)activism, characterised by Coleman (2014) as 'the weapons of the geek', all draw on elite technical know-how, further empowering the already empowered, but in different ways. Data activist strategies do not, therefore, appear to be the weapons of non-expert groups and citizens, despite Milan's (2014) claim that data activism is distinct from traditional activism in its ability to mobilise ordinary people.

Whether ordinary, not-already-expert groups and individuals are involved in data activism is something that I explored on a short data sprint at the DMI Winter School in January 2015, on a project entitled 'Mapping the data revolution', which focused on open data movements and was coordinated by Jonathan Gray of the Open Knowledge Foundation. Working with Christoph Raetzsch of the Freie Universität Berlin, Ivar Dusseljee and Jan-Japp Heine of the University of Amsterdam, we explored whether open data has traction outside what Gray called 'the open data bubble' (that is, people actively engaged in open data initiatives), by investigating who was talking about open data in online spaces. Focusing on the UK and using a dataset of tweets about open data provided to us by the DMI, we looked at the 24 top UK-based tweeters on the topic of open data (who had used the hashtag #opendata or the term 'open data'). We found that nine accounts belonged to individuals, six to businesses, six to NGOs and three to governmental bodies. While the highest number of tweeters were individuals, almost all work with or in data in NGOs, government and business, and all followed major open data initiatives like the Open Data Institute and the Open Knowledge Foundation Network—in this sense, they could be described as being 'inside' the open data bubble. Moreover, these nine tweeters, 37.5% of the cohort, had less than 12% of the total follower community for the 24 top tweeters. These results are shown in Figure 8.2.

In another experiment, we did co-hashtag analysis on the dataset we were given to identify the topics that are discussed alongside open data, and found that specialist terms like 'big data', 'open government' and 'open science' dominated, as shown in Figure 8.3. Terms which might be considered to represent ordinary concerns, such as safe streets, weather,

Followers per account type

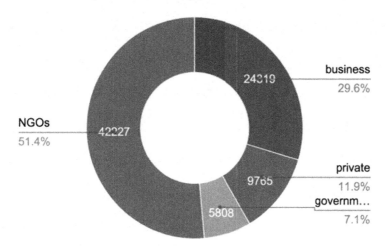

Figure 8.2 Findings from experiments into who tweets about open data

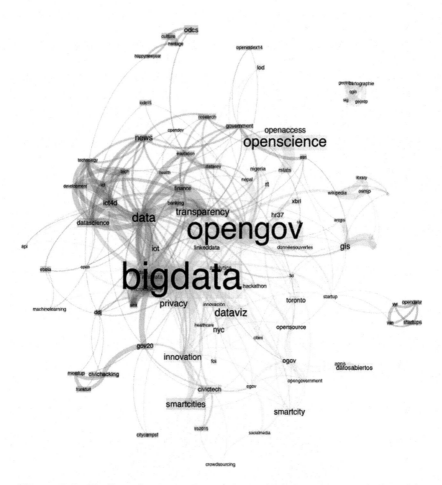

Figure 8.3   Findings from experiments into which terms are used alongside 'open data'/#opendata

health care, transport, were rarely used. We then carried out search engine link analysis to identify the extent of linkage to open datasets in Leeds, London and Manchester, and found only 24 links to London datasets, four links to Leeds datasets and two to Manchester datasets, suggesting very little active use of publicly available datasets. Numbers linking to national UK datasets varied, depending on the search engine used, from 21 to 57—still comparatively low numbers. As the research discussed in

Chapter 4 had identified that comment spaces on local newspapers' websites are important sources of debate on local topics, we searched local newspapers in the cities in which we had investigated dataset linkage, and found only one article mentioning open data, from Leeds, which was 2 years old and which had no comments. We started the process of searching voluntary organisation websites (which might also be seen as collectives representing the concerns of ordinary people) for mentions of open data, and found only one mention in 553 pages, after which we were forced to stop this experiment as our search was terminated by Google. We also did a LexisNexis search for use of the term 'open data' in mainstream UK newspapers and found only six articles since 1980, across two newspapers.[3]

These were modest experiments, undertaken by people not very expert (yet) in digital methods, using tools which delimit what can be researched and with which we were not previously familiar. Many of the experiments were undertaken on a dataset provided to us, so we had limited knowledge of the conditions of its generation. I do not adopt the kind of systematic approach to describing clearly the conditions under which the data were made that Driscoll and Walker advocate here, because I cannot, we were given the dataset with which we worked, we did not produce it and anyway I am not claiming that these experiments represent rigorous research or provide concrete evidence of phenomena. They were just experiments. What they showed was little engagement with open data beyond expert communities: top tweeters were inside the open data bubble, and specialist terms like 'big data', 'open government' and 'open science' were more commonly used alongside 'open data' than terms which might represent ordinary concerns. Numbers of links to datasets were low, and mentions of the term 'open data' in newspaper comments were rare. It may be that ordinary groups or citizens are engaging with open data in other contexts, or offline, or without using the term 'open data'—such activities cannot be identified with the methods we used. Nonetheless, we found little trace of ordinary engagements with open data in our experiments.

Other forms of data activism are problematic in different ways. In an article entitled 'Hack for good: speculative labor, app development and the burden of austerity', Melissa Gregg (2015) levels a number of criticisms at civic hackathons. First, she argues that such events (like the Code For meetings mentioned above), which are more and more widespread if not

---

[3] Some of these words and all of the figures are taken from a report produced from the sprint and co-written by me, Christoph Raetzsch, Ivar Dusseljee and Jan-Japp Heine.

ordinary, result from and respond to the conditions of austerity in which they proliferate. They are a 'patch for government' in times of efficiency cuts, because they aim to solve social problems formerly within the remit of states, freeing those who should be responsible for the social good from their obligations. Second, they serve as a way of normalising young people entering the tech workforce to enforced free and 'sacrificial' labour, argues Gregg, using a phrase from Andrew Ross's (2003) study of new media workers at the turn of the millennium, which intends to capture the long hours, deferred rewards and willing submission to the harsh conditions which characterise both types of work. They accustom young, would-be data workers to the gruelling working conditions that they can expect when they enter the data workforce—such as around the clock coding without sleep—and so to a blurring of what she has called elsewhere the 'work/ life ruse' (Gregg 2011). They are nonetheless an expected pre-condition to securing paid work and, as such, constitute a kind of compulsory voluntarism, suggests Gregg. Lilly Irani (2015b) argues that involvement in the hackathon process produces 'entrepreneurial citizens', aligned with Silicon Valley-influenced visions of what constitutes good worker subjectivity and reproductive of neoliberal work-based social orders.

These criticisms—that open data serves neoliberal agendas, that it and other forms of data activism are elite activities, empowering the already empowered and not ordinary people, that hackathons relieve governments of their responsibilities and normalise sacrificial working conditions among the future data workforce—suggest that we should be concerned about data activism. I agree with many of these criticisms and I certainly do not see activist practices through the lens of naïve optimism, as if they are unproblematic correctives to data power. But neither do I wish to dismiss them as embodying and embedded within the corrupting forces of capitalism, neoliberalism and austerity. I suggest that, despite their flaws and limitations, we might argue that they represent a 'politics of possibility' of the kind that Gibson-Graham witness in the alternative economic models that they write about in *A Postcapitalist Politics* (2006). Despite their limitations, forms of data activism seek to realise alternative ways of doing data mining, alternative data structures and alternatives to data power. Thinking in this way, we can argue that data activism is not solely problematic—it also forms part of a politics of possibility.

As should be clear by now, this book represents an attempt to bring together critical thought, as seen in the critiques of different forms of data activism discussed here, with the problem-solving ethos that underlies much data activism. Like data activism, some academic social media research and some of the public sector experiments with data mining discussed in earlier chapters are also underwritten by a commitment to problem-solving, whether problems relate to the inequalities and injustices brought about by data power, the requirement that publicly funded organisations do more with less or misunderstandings about what social data can be taken to represent. A central tenet of this book is that it is both possible and necessary to bring these two ways of thinking together—that is, thinking critically and thinking through a problem-solving lens. They are not mutually exclusive after all. In other words, the book asks how we might solve the problems of data power that critical thought helps us to identify, and it does this by attending to what should concern us about social media data mining and seeking to identify data practices about which we might feel hopeful. Some of the examples of data activism discussed here might give us cause for hope. After all, according to Stengers and Z, hope is the difference between probability and possibility (2003, quoted in Gibson-Graham 2006, p. 1).

CONCLUSION

This chapter provides a brief overview of efforts to do good with data in academic and activist contexts. From academic social media data mining which engages critically with the digital methods that it mobilises, to a variety of re-active and pro-active data activism practices, significant energy is being invested in these domains to try to ensure that humans do not have to submit to the harsh logic of data power. These forms of mining and analytics, of social media and other data, aim to enable citizens, publics and social groups and to repurpose data mining for the social good. Some of them, such as the academic practices with which I concluded the first section, use social media monitoring methods to do social media monitoring critique and, in so doing, they contribute to un-black-boxing data mining methods. All of the examples discussed here have problems and limitations, whether these relate to serving neoliberal agendas, empowering the already empowered, doing the work of states or accustoming the future data workforce to appalling and unacceptable working conditions. But they are not *only* problematic. While we can find ways of arguing that all of these practices play a role in sustaining structures of power, as I argued in relation to the public sector experiments with social media data mining discussed in Chapter 4, it would be empirically inaccurate to suggest that they *only* do this. I develop this point more fully in the final chapter.

In this chapter, an intriguing paradox has arisen in relation to data privacy and openness. On the one hand, in the first half, I suggested that more recognition is needed of the fact that social media data in the public domain feel intimate and personal to the people who put them there, who consequently have expectations about the flows of their data that are not being met when data miners gather and analyse them. In this sense, this public domain data is not really public. On the other hand, in the second half, I mapped out a range of initiatives which propose that data should be *more* public, and that opening up data to publics is a form of democratisation—although, of course, these groups rarely argue this in relation to social media data. So an argument has emerged which suggests that less openness *and* more openness are potentially empowering. The question this paradox raises, then, is whether both can be true and if they are, how decisions are made about which datasets it is empowering to open up and which it is empowering to keep closed. This in turn brings us back to the thorny issues of ethics and normative judgements about data mining which emerged in this and previous chapters. These are issues that need addressing now, I have suggested, not at some point in the future, because of the new data relations that the logics of social media data mining and datafication usher forth, and the new ethical arrangements that these relations subsequently require. And this, in turn, brings us back to the suggestion I made at the end of the last chapter, that philosophically informed social theory can contribute to understanding these new data relations and developing these new ethical arrangements. I say more about this, too, in the next chapter.

Can the data practices discussed in this chapter be considered ordinary? Academics and data activists are elites in different ways, but, as suggested in the introduction, the proliferation of data mining in these spheres might be seen as entangled with the becoming-ordinary of social media data mining. Although the primary focus of this book is on data mining in ordinary *organisations*, not among ordinary *people*, data activism nonetheless opens up a range of questions about ordinary people's engagements with data. For example, are the people uploading and sharing content on platforms like Ushahidi (http://www.ushahidi.com/)—initially developed as an incident-reporting platform during post-election crises in Kenya in 2007 and since used for crisis data sharing in Haiti, Chile and elsewhere—data activists, or would we want to reserve that category for the people who built and manage it? Should we welcome the ways in which putting 'data' before 'activism' challenges traditional ideas about what constitutes activism, about who is inside and outside this category? The emergence and expansion of data activism requires us to rethink how and why we draw boundaries around categories, and unsettling the category 'activist' could be seen as a good thing—it too might open up a space for a 'politics of possibility'.[4] This issue of ordinary people's engagements with data is something else to which I return in the next, concluding chapter.

[4] These ideas developed in conversation with Noortje Marres.

CHAPTER 9

New Data Relations and the Desire for Numbers

This concluding chapter returns to the questions that have framed the rest of the book and offers some tentative answers to them. What should concern us about social media data mining? What established problems endure and what new problems surface? Must humans submit to the harsh logic of data power, or can the technologies, techniques and methods of data mining be repurposed, so that they can be used in ways that enable citizens and publics and that make a positive contribution to social life? Is there scope for agency within data mining structures and, if so, what does it look like? I start the chapter by returning to the critiques that I discussed in Chapter 3, which emerged at a time when data mining was primarily undertaken by major players like governments, security agencies and the social media platforms themselves, and I assess whether they apply to ordinary social media data mining. I then review the concerns that surfaced across the sites of my research in ordinary organisations doing data mining. I also return to some of the efforts to do good with data that I have discussed throughout the book. By noting these, I highlight that the becoming-ordinary of data mining does not only usher forth concerns; the prevalence of data also opens up opportunities for doing good with them. Together, these concerns and efforts to do good constitute new data relations, brought into being by patterns of datafication, the expansion of social media logic and the becoming-ordinary of social media data mining. I discuss the implications of these new data relations and then finish the chapter by pointing to some of the ways in which gaps in this book could be addressed by future research.

© The Editor(s) (if applicable) and The Author(s) 2016   221
H. Kennedy, *Post, Mine, Repeat*,
DOI 10.1057/978-1-137-35398-6_9

ESTABLISHED CONCERNS

In Chapter 3, I proposed that dominant criticisms of social media (and other) data mining could be summarised like this: that it results in less privacy and more surveillance; that it is mobilised for the purposes of discrimination and control; that access to it is unequal and this results in inequality; and that it should concern us methodologically, because the data it generates shape the social world in opaque and black-boxed ways. I expected that these dominant ways of framing data mining would not capture the practices of actors in small-scale organisations, and in many ways they do not. Some of the activities undertaken in these contexts are less surveillant, discriminatory and privacy-invasive than those undertaken by large-scale actors. I did not see much evidence of data mining being used for discriminatory purposes in my research—although of course it is possible that some discrimination occurred in applications of mined data of which I was not aware. As social media and other forms of data mining become ordinary, the actors engaged in it and their aims and purposes diversify, and these ordinary data mining practices do not generate the same concerns as extraordinary forms.

But being ordinary does not make data mining beyond critical interrogation. Surveillance, discrimination and privacy-invasion occur at the level of the ordinary as well as the extraordinary. Turow et al. (2015b) argue that ordinary and everyday practices like geofencing in retail spaces—that is, the use of GPS (Global Positioning System) or RFID (Radio-frequency Identification) to identify the geographical proximity of a shopper to a store and then to remind the shopper of the discounts he or she has been offered—represents a form of ordinary, everyday surveillance, and of discrimination too, as not all shoppers' offers are equal. And, as a number of other writers have pointed out, such as Andrejevic and Gates (2014), McQuillan (2015) and van Dijck (2014), in times of datafication, the wholesale mining of all data about everyone and everything, all of the time, has become ordinary. We have seen some evidence of this in the organisations which recruit the services of social insights companies, discussed in Chapter 6, in which the very data mining tools and techniques that workers are required to use are also utilised to monitor and evaluate worker performance.

We have also seen that privacy—and its entanglements with the ethics of data mining—matter, even though these terms, privacy and ethics, are not necessarily the most helpful for capturing what is at stake. In the focus groups that I discussed in Chapter 7, participants' discomfort with the mining of what feels to them to be private data reflects their concerns about whether data miners are operating ethically. In that chapter, I noted that even though the focus groups were not framed in terms of privacy, at times participants brought the discussion round to this topic. Participants suggested that social media *feel* private, personal and intimate, even when they are not. I argued in that chapter that the fairness of data mining practices matters to social media users—and I come back to that later—but privacy matters too. Data miners or organisational actors using the services of data intermediaries sometimes acknowledge this as well. In Chapter 5, I noted that a number of interviewees in intermediary social media data mining companies recognised that 'public' is not a straightforward category in social media contexts and in Chapter 6, we also saw some recognition of this, for example by Jane, the Head of Digital Communications in one of the universities. In these ways then, and as boyd (2014) notes, the issue of privacy has not gone away.

Neither have access and inequality. Some of the research discussed in the previous chapters aimed specifically to address these issues. The action research discussed in Chapter 4 explored whether it was possible for those with limited economic means to access and use social media data mining and some of the examples of data activism discussed in Chapter 8 are motivated by similar objectives. Inequalities endure, and so do efforts to address them. Likewise, epistemological concerns about social media data mining do not go away when these methods are in the hands of ordinary organisations. They still work in the same way, they still have effects, make differences and enact realities (Law and Urry 2004), and raw data is still an oxymoron (Bowker 2005; Gitelman and Jackson 2013). I return to these methodological issues in the next section.

Although we need to think differently about data mining when it is in the hands of the not-powerful and we cannot make the same overarching criticisms about surveillance, control and inequality that apply to large-scale data mining, actually, some of these established concerns do still matter, despite the mundane and small-scale character of the data mining activities and the adjustments that result. What's more, new issues emerge through the normalisation of data mining. These include the prevalence of a desire for numbers and its various troubling consequences, such as its effects on critical thinking about data-making and on work and workers. I discuss each of these issues below, and then return to the question of whether it is possible to do good with data.

EMERGING CONCERNS

The Desire for Numbers

One significant issue that emerged across the sites of my research is the desire for numbers that engaging in social media data mining in the broader context of data power elicits. Earlier in the book, I suggested that this notion of a desire for numbers might be seen as a convergence of Porter's (1995) ideas from the mid-1990s about trust in numbers, and Grosser's (2014) more recent concept of a 'desire for more', developed in a context in which the metrification of sociality, primarily on social media platforms, creates a desire for ever more metrics. Merging these ideas helps us account for the hunger for data and statistics that I identified *despite* knowledge about their inaccuracy and unreliability, *despite* distrust. The datafication of the ordinary and the everyday makes this a widespread phenomenon—what was once qualitative is now measured quantitatively, so quantities are desired in relation to more things, to that which was previously qualitative. Enthusiastic responses to reports we produced for public sector organisations, discussed in Chapter 4, anecdotes about how the inaccuracy of data was irrelevant to clients of intermediary insights firms in Chapter 5, and stories about 'the fetishism of the 1000' from some of these client organisations which formed the subject matter for Chapter 6 all serve as examples of this widespread desire for numbers.

In those chapters, I argued that the strength of this desire for numbers makes it difficult to open up spaces to talk about the limitations of data mining and the data it produces. In Chapter 4, the combination of data mining methods and action research worked together to limit the research team's success in achieving our original intent of talking about the problems of social media data mining with our partners, as well as its potential. Doing data mining was motivated by a desire to produce results; data mining produces the expectation that data would be found. We also saw this desire for numbers in the interviews we undertook prior to the action research. Quantitative data, produced through systems like Museum Analytics, were desired by managers and funders, with no apparent concrete consequence. The 'data gathered' box was ticked, the desire for numbers was fulfilled and data were filed away.

In Chapter 5, there was further evidence of this desire for numbers, this time in the context of the work of commercial social insights companies. Here, we saw how this desire suppressed discussion between workers in these companies and their clients about the limitations of data mining. It also led to a lack of interest in precise numbers—sometimes interviewees felt that clients would be satisfied with any number, however inaccurate. Some interviewees expressed frustration at this—they would have liked to talk to clients about the challenge of obtaining good quality, accurate social data and about what the numbers that social insights produce actually represent. These insights professionals were alert to the inadequacies of the numbers they produced, even as they produced them. In the organisations which use the services of insights companies discussed in Chapter 6, the desire for numbers came fully into view. This could be seen in the evangelism of people within organisations who undertake social media insights (which sometimes leads to an overstatement of the changes that result), their own and their colleagues' faith in what metrics can do, and their frustration when expectations were not met. Academic and activist data miners are not immune to this desire either, even as they seek to repurpose the tools of data mining, unveil its mechanics or apply it for the social good.

The use of social media data mining is motivated by, produces and reproduces a desire for numbers which already exists in a context of datafication and social media logic. In this context, numbers are assigned new powers—this results in a belief that the quantitative is all we need, that the numbers will speak for themselves. And this in turn means there is a risk of devaluing the qualitative. As noted in previous chapters, Porter argues that numbers are for managing the world, not understanding it. So devaluing the qualitative can mean a loss of understanding; this is why Baym (2013) states that we urgently need 'qualitative sensibilities and methods to help us see what numbers cannot'. Even when there is awareness of the limits of numbers, once they are 'in the wild', they become separated from knowledge about what they can be taken to represent, as I suggested in Chapter 3 (Gerlitz 2015, citing Espeland and Sauder 2007). This was the case in the research I undertook: reflection on the numbers that were being produced that might produce understanding of what they represented was lost or forsaken because of this desire. In the next two sections, I say more about what is lost when a desire for numbers prevails.

(Not) Thinking Critically About Data-making

This section reflects on the possibility of thinking critically about data mining (or moving ideas that might be commonplace in the social sciences into other spaces in which data mining takes place) in the context of a desire for numbers. In Chapter 3, I quote boyd and Crawford's criticism of the belief that 'large data sets offer a higher form of intelligence and knowledge that can generate insights that were previously impossible, with the aura of truth, objectivity, and accuracy' (Boyd and Crawford 2012, p. 663). I highlighted the range of actors, decisions and processes that shape data and data mining, and the authors who have drawn attention to the various ways in which data mining methods make and shape the data they produce. As social media data mining becomes more ordinary and widespread, to what extent is there evidence of such epistemological understanding outside academic settings? I found some traces of it in my research. In my interactions with public sector workers, participants showed awareness that the data produced through these methods are not representative of whole publics; this meant that they were unlikely to base important decisions on the findings of such methods alone. However, although the research team involved in this project was aware of how various aspects of the data mining process (like keyword choice) were shaping the things they aimed to unveil, the power of the desire for numbers left little room for discussion of these matters with our partners, as noted above. Critical discussion of these epistemological issues in relation to data mining did not take place with partners; they remained the concern of the academic research team.

In interviews in organisations which contract the services of insights companies (or do their own social analytics), there was more awareness of the problems with data mining and what the data these methods produce can be taken to mean. This is not surprising, as these interviewees were more experienced in social insights than the public sector workers who participated in our experimental action research. It was not surprising that these interviewees showed some understanding of the difficulties of working with social media data. Some recognised technological problems in doing data mining, such as tools' limited ability to capture certain kinds of information. Others noted that there is too much social media data, that such data are unstructured, badly formatted and therefore difficult to analyse. Social media data are unreliable because users may have multiple accounts, or hard to assess comparatively because of constant changes in platforms, APIs and algorithms, interviewees noted. The most sophisticated understanding of these issues could be seen, again not surprisingly, among professionals working in social insights companies. Interviewees talked about how their companies grapple with the poor quality of social data, and expressed frustration at clients' lack of attentiveness to the limitations of the data that they shared with them. Some noted that who is active on social media gives an inaccurate impression of who constitutes a brand's audiences. Others acknowledged that they manipulate and filter data to get 'exactly what the brand wants', that data are not simply found, but made and shaped. For all of these reasons, some of these interviewees described social media data mining methods as blunt.

This recognition of the limitations of social media data and data mining might give us some cause to be hopeful—there was little trace of an unquestioning faith in large-scale, digital data of the kind that boyd and Crawford criticise across the sites of my research. But these are issues that research participants acknowledged when asked about them, or when pushed. Or, indeed, some of them I identify as I peer through the cracks of participants' corporate 'there are no problems here' responses to my questions about what might concern us about data mining. These are not the subject of widespread conversation among people engaged in ordinary data mining. As noted above, the dominant desire for numbers overshadows such conversations, and this desire is the result of a broader culture of datafication and social media logic, which is difficult to resist. What's more, despite acknowledgement of the limitations of social media data mining, faith in their abilities endured, as seen in the data evangelism of the interviewees discussed in Chapter 6, the ideas about its potential uses among action research participants in Chapter 4, and elsewhere. The desire to do good with data of activists, academics and some public sector workers also reveals a faith in their potential. The many limitations of data mining that participants recognised seem to lead to an obvious conclusion: that it is not really possible to know the social world with these methods. Yet there is little evidence of those involved in data mining arriving at this conclusion. This is understandable, as they have commercial imperatives to continue to mine data. But for those of us who would like to see more public reflection on the ways in which data are made and shaped, this remains a major challenge.

Another thorny aspect of (not) thinking critically about data-making relates to the view that social media data are public and therefore fair game to be mined and analysed by interested parties. The absence of discussion of how citizens and publics feel about having their data mined in the public sector action research might be seen as evidence of this. Whether organisations like councils and museums should be transparent about their data mining activities, and whether social data should be considered public or private, were not on the agenda in the conversations we had with them. Although one of the commercial company interviewees said that in 'public sector organisations there's almost a reticence, are we allowed to have this type of information?', I saw no evidence of this. And in these commercial companies, despite acknowledgement that 'public' is not a straightforward category in social media contexts, most continue to adhere to the 'it's public so it's fair game' mantra. Among clients of these commercial companies, questions relating to the publicness or privateness of social media data were only discussed when brought up by the interviewer, and when they were, respondents also adhered to this belief. Some interviewees thought that the public service remit of their organisations made their data mining ethically acceptable and so excused them from the need to reflect on these issues.

This view that social data are 'public so fair game' stands in contrast to social media users' reflective assessments of when they consider it to be acceptable for their social data to be mined and when they do not in the focus groups discussed in Chapter 7. Among these users, ideas about *fairness* had stronger traction than the idea that social data are 'fair game'; in the view of some of them, social data are neither straightforwardly public, nor are they fair game for mining. But among data miners, despite recognition that the interests of social media users might differ from those of social media analysts, as with other issues, these concerns were put aside in order to fulfil the desire for numbers, their own, their clients' or their superiors'.

Work Effects

The spread of social media data mining and the desire for numbers that accompanies it has effects on workers, and this is another concern that emerges as social media data mining becomes ordinary. Data mining is changing people's working conditions across a range of sectors. This was evident in the public sector organisations discussed in Chapter 4, where marketing, communications, engagement and other teams in local city councils and museums are starting to think about how and in what ways to integrate social insights into their work. They are attempting to do so in and because of difficult conditions—cash-strapped and under-resourced, data mining is seen as a potentially cheap and efficient way of understanding and engaging publics. At the same time, this under-resourcing makes it difficult for public sector workers to invest time to develop the expertise required to do data mining well.

Work issues were most visible in the organisations which engage the services of social insights companies discussed in Chapter 6. Here, modifications to working arrangements made as a result of internal, organisational shifts to accommodate data mining were seen to have consequences for a range of workers within the organisations. Roles undertaken expand and proliferate and workers are expected to keep up to date with new technical developments. As a consequence, boundaries around working hours dissolve to make way for engaging on social media at optimum times, and, as noted earlier in this chapter, the social media data mining methods that workers are required to use are turned back on them, as their social media performance is monitored and evaluated on social media. In these ways, we can see a troubling relationship between the becoming-ordinary of social media data mining, the desire for numbers that this generates, and changes to the working lives of employees in ordinary organisations. In the broader context of work/technology relationships, this is not new, of course—many writers before now have highlighted how technologies intended to make work more efficient have in fact increased the burden of work for employees in organisations (for example Webster 1999).

There are other ways in which the proliferation of data and data mining, and the desires they produce, impact on work. This was noted in relation to some of the alternative data practices that I discussed in Chapter 8. There, I pointed out that some writers have argued that data activism events like civic hackathons normalise precarious working conditions, as participants work around the clock to solve social problems (Gregg 2015). They produce 'entrepreneurial citizens' (Irani 2015b) aligned with particular, Silicon-Valley-type visions of what constitutes good worker subjectivity. Hackathon participation might be seen as a new form of free and sacrificial labour, a kind of compulsory voluntarism, accustomising young, would-be data workers to gruelling working conditions and so to a blurring of the 'work/life ruse' (Gregg 2011), simultaneously reproductive of neoliberal work-based social orders. These too are some of the work-based consequences of desiring numbers.

Chapter 5 focused on work too, but in a different way. Whereas the examples given above might be said to reflect a deterioration of agency for workers (they have less control over when they work and the boundaries around their workloads, and less autonomy from the employer's surveillant eye), that chapter highlighted the role of ethical considerations in work-based, data-driven decision-making processes in intermediary, commercial social media insights companies. In this way, it pointed towards micro-level acts of individual agency in the context of data mining. I showed how individual ethical barometers influenced workers to draw lines around what they will and will not do, and that these lines are not in the same place for all workers. Ethical considerations influence the decisions that are made by social media data miners, but such decisions are not only ethically informed—they are also economic. These examples show that workers in this sector act with agency in their decision-making, they do not simply submit to data power's logic. But they do this within the structures of data mining—the ground does not shift much, and the structures remain intact. I return to what this tells us about the possibility of agency after a note about doing good with data.

DOING GOOD WITH DATA

So these are the things that might concern us about social media data mining becoming ordinary: widespread datafication that produces a persistent desire for numbers, which makes it difficult to create space for reflection about how data are shaped by the conditions in which they are produced; a pervasive belief that social media data are fair game for mining because they exist in public space, even when there is acknowledgement that conventional notions of what constitutes the public and the private do not apply in social media spaces; and the various troubling effects on work and workers of the take up of data mining across ordinary organisations. My research shows the prevalence of these troubling phenomena, how they materialise in specific contexts and what it is like to live and work with them.

But I do not think that we should *only* be troubled by ordinary social media data mining. The spread of data and of data mining also opens up the possibility of doing good things with them, as seen in my discussion of public sector experiments, alternative data practices and critical academic research. Local councils and museums using data mining to understand how their publics are networked to each other, to identify influential individuals or groups with whom to engage or the most beneficial social media channels to use in order to engage their publics might be in the public interest. This is certainly how actors experimenting with these methods see it. Pro-active data activism, such as open data movements, citizen science projects, initiatives aimed at linking up data science expertise with social needs or facilitating the mining of one's own data, civic hackathons, and the work of data artists, data visualisers, tricksters, hoaxers and other actors-from-the-outside, sit alongside critical social research with digital methods, in attempting to repurpose data mining for different kinds of social and public good.

As noted in an earlier chapter, concepts like doing good, engaging and empowering are not straightforward. They play a role in disciplining subjects and regulating citizens, however well-intentioned. Open data initiatives draw on elite technical know-how and, as such, are not populated by ordinary, non-expert citizens. Open data strategies are mobilised in the interests of neoliberal political agendas and empower the already empowered. Hackathons normalise precarious and 'sacrificial' labour and produce entrepreneurial subjects. Some academic social media data mining is complicit in the normalisation of datafication, dataism and dataveillance (van Dijck 2014). None of the data mining practices which aim to 'do good' are without problems, nor do they simply subvert data power for the social good. However, as also noted, although public bodies, open data movements, civic hackers and others' efforts to do good with data can all be read as complicit with apparatuses of control, these practices need to be seen through more than *just* a critical lens, as not *only* embodying and embedded within the corrupting forces of capitalism. To apply only a critical lens would be empirically inaccurate, I have argued.

Efforts to do good with data can be seen as efforts to *problem-solve* with data. I argue that when it comes to analysing the structures of data power, and considering whether it is possible to act ethically within them, we should bring critical thought and problem-solving together. They are not mutually exclusive. And yet, critical thinkers sometimes construct them in this way, and the problem-solving ethos of some of these initiatives is dismissed in an effort to unveil their underlying complicity with structures of power. But while we can find ways of arguing that all of these practices have their place in the capitalist techno-social complex, this is not the only way of seeing them. Thinking with *both* critical *and* problem-solving perspectives is essential, I suggest, for addressing the questions of what should concern us about social media data mining and whether there are forms of data mining that can be enabling of the non-powerful. Returning to Andrew Ross's book about new media work (Ross 2003), from which Gregg (2015) takes the notion of sacrificial labour, we find other ideas that we can apply to efforts to do good with data. In that book, Ross argues that we need to do more than just account for the problems of new media's no collar work and its hidden costs. This one-dimensional perspective does help us understand the pleasures, passions and possibilities that such work is seen to offer up by the people engaged in it, he proposes. The same multi-dimensional approach is needed in relation to social media data mining—to account for its problems, but also its potential.

NEW DATA RELATIONS

As social media data mining becomes more and more ordinary, and as data-fication and social media logic take hold, new data relations emerge, which are increasingly integral to everyday social relations. As we post, mine and repeat, these new data relations bring with them new questions about ethics, agency and social life. All of these things are, at the time of writing, unstable—social media data mining assemblages are under-determined, argue Marres and Gerlitz (2015), 'multifarious instruments', says Marres (2012), objects with which to experiment. Some of the ambivalences that we have seen throughout this book and to which I have returned in this chapter demonstrate this instability—about ethics, privacy, trust and distrust in relation to data mining. In Chapter 6, I introduced the concept of interpretative flexibility to characterise the ethical ambivalence of social media data miners, a term used in Science and Technology Studies to describe socio-technical assemblages for which a range of meanings exist, whose definitions are as yet undetermined. We might see all of the things noted here—data, data mining, ethics, privacy, agency, trust and distrust—as being in a state of interpretative flexibility.

For example, efforts to do good with data could be seen as evidence of the possibility of individual and collective agency within data mining structures, or of acting ethically, to return to the definition of agency that I use in this book. The prevalence of data mining makes it possible to experiment with re-designing data mining to better serve publics. We saw some worker agency too, as noted in this chapter—among professional insights workers, deciding what data mining work they will and will not do. In these senses, in answer to Feenberg's adapted question about whether we must submit to data mining structures, we might respond 'no'. But acts of individual or small group agency do not mean that structures have been torn down, that democratisation has been established, or that the not-powerful have gained power. For we have also seen an absence of agency through the work effects I have traced—with no control over working hours and workloads, and being subjected to social media monitoring themselves, some workers lack agency, despite their enthusiasm for data mining. This uncertainty with regard to agency represents one example of the interpretative flexibility of the new data relations that emerge in times of ordinary social media data mining.

In this context in which new concerns and old concerns merge and new possibilities arise, new ethical questions also surface. These too are unstable. Are social media data that are in the public domain public or private? The answer is not clear. Because of this, we have seen a lack of certainty about how to be, ethically, among participants, for example as they acknowledge some of the ethical complexities of data mining but do it anyway. Also because of this uncertainty, the individual decisions of data workers shape how data mining gets done and how it gets stabilised. Interviewees' histories inform their decisions, from their educational backgrounds to their experiences as workers, volunteers, and human subjects. But so do company policy and the fact that it is their job to contribute towards fulfilling the desire for numbers. So, despite acknowledgement that 'public' is not a straightforward category in social media, most respondents adhere to the mantra that if social media data are public they are therefore 'fair game' to be mined and analysed. This is further evidence of instability of these new data relations. But whereas Marres and Gerlitz argue that we should suspend normative judgement while we experiment with this under-determinacy, I argue that both are possible and desirable: we need to experiment with data mining's potential to do good, but we also need to interrogate it normatively. I come back to how we might do this at the end of this chapter.

Social media are at the heart of the new data relations. The fact that social media data are technically more available to mine than other types of data is central to the emergence of these new relations—more social media data means more data mining, which means more numbers and a greater desire for numbers. The effects on work of more and more data mining are also fuelled by the widespread availability of social media data in particular. And some of the instability discussed here relates specifically to social media data mining. The unresolved matter of what is public and what is private and, relatedly, what it is fair to mine and what it is not, is specific to social media. Because of the private and intimate character of social media data, it is hard to envisage whether the practices which aim to do good with data, to open them up to wider access and use, could work in relation to social data; the experiments on the Our Data, Ourselves project discussed in Chapter 8 remain inconclusive. While some doing good with data is under way, doing good with social media data is harder to imagine—this uncertainty too is integral to the new, emerging and unstable data relations that I have traced in this book.

DOING *BETTER* WITH DATA

In this final section, I point to three absences in this book and two areas for further research, which might help to answer the question of how to do *better* with data. The first absence relates to regulation, and I do not suggest that it requires further research, because research in this area is well under way. Clearly, we need to consider what kinds of structures need to be in place in order to facilitate better data relations and to make it possible to do better with data—this was evident in discussion with commercial social media analysts and with social media users. How to regulate data practices is an important question, but it is not one that I have addressed in this book, because I am not an internet regulation expert and that work is best left to people who are. There is plenty of great regulation research, lobbying and activism, but I have not discussed it here, because I do not know it well, and that would be another book. What would be helpful, I suggest, is to bring research about regulation into dialogue with research about experiences of data mining discussed in this book and with the other two areas mentioned below, as this might inform better data regulation.

The first area for further research relates to data work, the people involved in it and their processes, of which we need more understanding. Some scholars have attended to the work of the data scientist (for example Gehl 2014, 2015; MacKenzie 2013), but there are many more roles involved in the process of producing data than this. Data cleaners, algorithm writers and other statisticians, data visualisers, designers of the interfaces of systems that gather and output data are just a few of them. Studying these workers will help to address questions of where power lies in data-making. Often, digital workers are held responsible for the systems that they contribute to produce, as if they were all-powerful—this is the case in Adam and Kreps discussion of web design for accessibility (Adam and Kreps 2006) and can also be seen in Munson's work on the designers of recommendation systems (Munson 2014). But this is not the case. Power does not operate in simplistic ways and the location of power in data-making processes is more complex than this. This is why we need more understanding, through studies of data workers, of how data and their representations come into being.

Second, we need more understanding of ordinary people's relationships with data. This is an issue of user agency, something I have not really addressed in this book, as I have focused on data mining in ordinary organisations, not by ordinary people, and so on worker and techno-agency. But as data acquire new powers, it is important that we understand user agency in relation to data structures, and that ordinary people understand what happens to their data, the consequences of analysis of their data and the ways in which data-driven operations affect us all, in order to be able to participate in datafied social, political, cultural and civic life. We need to think about whether it is possible for ordinary people to do the same things with their data that corporations and organisations can do. The Quantified Self movement is one example of a field in which individuals attempt to take ownership of their own data (Nafus and Sherman 2014), although critics point out that corporations ultimately benefit from these data gathering practices (for example Crawford et al. 2015). Another example of an attempt to develop understanding of ordinary people's engagements with data is the Seeing Data project (http://seeingdata.org/) on which I have been working with William Allen, Rosemary Lucy Hill and Andy Kirk, which explores how people engage with data through visualisations. These examples represent the beginnings of thinking about ordinary people's engagements with data.

To address these important questions of how people engage with data, a number of approaches are possible. We could look at them from the perspective of data literacy, as Pybus et al. (2015) do in their project about big social data. Addressing data literacy requirements means thinking about how we learn to relate to numbers and statistics, and this in turn means thinking about whether and how data matter to people. This brings us back to some of the ideas that I introduced in Chapter 7, in my discussion of social media users' concerns about the fairness of social media data mining practices. In that chapter, I suggested that future research into data mining would benefit from further engagement with concepts like well-being and social justice, in order to consider whether a better relationship between social media data mining and social life is possible. As Sayer (2011) notes, concepts like flourishing and well-being are unfamiliar territory for social scientists, and yet they cannot be avoided if we are to attempt to understand how greater social justice might be achieved, in relation to data mining as well as in other spheres. I cannot see how we can address issues of why, how and whether data matter to people without turning to these concepts. Doing so will also help to address some of the ethical questions that new data relations bring to the fore, about how evaluations of fairness might guide data mining practices and who gets to control what definitions of fairness count. These are all tricky, philosophical questions, and I am no philosopher, so I have simply hinted at these issues here. There is much more to be done.

It is possible to overstate the importance of data and data mining, to ordinary people, to workers in ordinary organisations, and in society more generally. Singling them out as objects of study and talking to people who are engaging with them, we can overlook the possibility that people might feel that data and their mining do not matter. Asking people engaged in data work to talk about numbers and then concluding that numbers matter could seem somewhat tautological, I admit. But they do matter, because datafication, social media data mining and social media logic are pervasive and enduring. A qualitative book about quantities, like this one, cannot comment categorically on how widely they matter, or the range and extent of some of the things I have discussed, of course—more mapping is needed, across the domains discussed here, and others as they emerge. This will also help improve understanding of how to address and confront those things that concern us about social media data mining and how to do it better. But we also need to remember Baym's (2013) assertion that qualitative sensibilities are needed in times of datafication 'to help us see what numbers cannot'. Small-scale research about methods for engaging with large-scale social data remains important. Understanding qualitatively the varied and specific ways in which data mining is enacted in particular contexts, the possibilities it opens up and the problems it ushers forth will help us move towards what feels like the right balance of empirical accuracy, critical thought and problem-solving.

BIBLIOGRAPHY

Adam, A., & Kreps, D. (2006). Enabling or disabling technologies? A critical approach to web accessibility. *Information Technology and People*, 19(3), 203–218.

Albrechtslund, A. (2008). Online social networking and participatory surveillance. *First Monday*, 13(3). Available at http://firstmonday.org/article/view/2142/1949. Accessed 10 Jan 2014.

Anderson, C. (2008). The end of theory: Will the data deluge make the scientific method obsolete? *Edge*. Available at: http://www.edge.org/3rd_culture/anderson08/anderson08_index.html. Accessed 7 May 2010.

Andrejevic, M. (2004). *Reality TV: The work of being watched*. Lanham, MD: Rowman & Littlefield Publishers.

Andrejevic, M. (2007). *iSpy: Surveillance and power in the interactive era*. Lawrence, KS: University of Kansas Press.

Andrejevic, M. (2011). The work that affective economics does. *Cultural Studies*, 25(4-5), 604–620.

Andrejevic, M. (2013). *Infoglut: How too much information is changing the way we think and know*. New York: Routledge.

Andrejevic, M., & Gates, K. (2014). Big data surveillance: Introduction. *Surveillance and Society*, 12(2), 185–196.

Anonyzious (2012). 10 largest databases of the world. *realitypod.com*. Retrieved 24 March 2012, from http://realitypod.com/2012/03/10-largest-databases-of-the-world/. Accessed 16 Oct 2014.

Anstead, N., & O'Loughlin, B. (2014). Social media analysis and public opinion: The 2010 UK general election. *Journal of Computer-Mediated Communication*, 20(2), 204–220.

© The Editor(s) (if applicable) and The Author(s) 2016   237
H. Kennedy, *Post, Mine, Repeat*,
DOI 10.1057/978-1-137-35398-6

AOIR (2012). *Ethical guidelines.* Available at: http://www.aoir.org/reports/ethics2.pdf. Accessed 27 Feb 2013.

Arnott C (2012). *Internet privacy research.* Available at: http://cccs.uq.edu.au/documents/survey-results.pdf. Accessed 16 July 2013.

Arvidsson, A. (2011). General sentiment: How value and affect converge in the information economy. *Social Science Research Network.* Available at: http://papers.ssrn.com/sol3/papers.cfm?abstract_id=1815031. Accessed 4 May 2011.

Arvidsson, A., & Peitersen, N. (2009). *The ethical economy.* Available at: http://www.ethicaleconomy.com/info/book. Accessed 1 Mar 2010.

Attewell, P., & Monaghan, D. (2015). *Data mining for the social sciences: An introduction.* Berkeley, CA: University of California Press.

Baack, S. (2015). Datafication and empowerment: How the open data movement re-articulates notions of democracy, participation and journalism, *Big Data and Society,* (forthcoming).

Ball, K. (n.d.). *The Surprise Project* (surveillance, privacy and security). Available at: http://surprise-project.eu/. Accessed 23 Mar 2015.

Banks, M. (2007). *The politics of cultural work.* Basingstoke, England: Palgrave Macmillan.

Barnes, T. (2013). Big data, little history. *Dialogues in Human Geography,* 3(3), 297–302.

Barnes, M., Newman, J., & Sullivan, H. C. (2007). *Power, participation and political renewal: Case studies in public participation.* Bristol: Policy Press.

Barocas, S., & Selbst, A. (2014). Big data's disparate impact. *Social Science Research Network.* Available at: http://papers.ssrn.com/sol3/papers.cfm?abstract_id=2477899. Accessed 2 Oct 2014.

Bartlett, J. (2012). *The data dialogue.* Available at: http://www.demos.co.uk/publications/thedatadialogue. Accessed 16 July 2013.

Bassett, C. (2015). Plenty as a response to austerity? Big data expertise, cultures and communities. *European Journal of Cultural Studies,* 18(4-5), 548–563.

Bates, J. (2013). Information policy and the crises of neoliberalism: The case of Open Government Data in the UK, *Proceedings of the IAMCR 2013 Conference, 25-29 June 2013.* Available at: http://eprints.whiterose.ac.uk/77655/7/WRR0_77655.pdf. Accessed 29 Sept 2014.

Baym, N. (2013). Data not seen: The uses and shortcomings of social media metrics, *First Monday,* 18(10). Available at: http://firstmonday.org/ojs/index.php/fm/article/view/4873/3752. Accessed 11 Feb 2014.

Baym, N., & Boyd, D. (2010). Socially mediated publicness: An introduction. *Journal of Broadcasting and Electronic Media,* 56(3), 320–329.

BBC (2014). Facebook sues fake "like" scammers for £1.3bn, *BBC News Online.* Available at: http://www.bbc.co.uk/news/29505104#utm_sguid=142547,53996041-96f0-dbc6-5a1b-de827661ffcb. Accessed 9 Oct 2014.

Beer, D. (2009). Power through the algorithm? Participatory web cultures and the technological unconscious. *New Media & Society,* 11, 985–1002.

Beer, D., & Burrows, R. (2013). Popular culture, digital archives and the new social life of data. *Theory Culture & Society,* 30(4), 47–71.

Belk, R. (2010). Sharing. *Journal of Consumer Research,* 36(5), 715–734.

Benkler, Y. (2006). *The wealth of networks: How social production transforms markets and freedom.* New Haven, CT: Yale University Press.

Bkaskar, R. (1979). *The possibility of naturalism: A philosophical critique of the contemporary human sciences.* Brighton, England: Harvester Press.

Boellstorff, T. (2013). Making big data, in theory. *First Monday.* Available at: http://firstmonday.org/ojs/index.php/fm/article/view/4869. Accessed 11 Feb 2014.

Bollier, D. (2010). *The Promise and Peril of Big Data.* Available at: http://www.aspeninstitute.org/publications/promise-peril-big-data. Accessed 4 June 2012.

Borsboom, B., van Amstel, B., & Groeneveld, F. (2010). Please Rob Me website. Available at: http://pleaserobme.com/. Accessed 4 Dec 2011.

Bounegru, L., de Gaetano, C., & Gray, J. (2015). Open data in mainstream media, part of DMI data sprint 'Mapping the open data revolution', Digital Methods Initiative Winter School, University of Amsterdam, 16 Jan 2015.

Bourdieu, P. (1980). *The logic of practice.* Stanford, CA: Stanford University Press.

Bowker, G. (2005). *Memory practices in the sciences.* Cambridge, MA: MIT Press.

Boyd, D. (2010). Making sense of privacy and publicity. Keynote Speech, SXSWi (South by South West Interactive), Austin, TX, Retrieved 13 March 2010, from http://www.danah.org/papers/talks/2010/SXSW2010.html. Accessed 20 May 2010.

Boyd, D. (2014). *It's complicated: The social lives of networked teens.* New Haven, CT: Yale University Press.

Boyd, D., & Crawford, K. (2012). Critical questions for big data: Provocations for a cultural, technological and scholarly phenomenon. *Information, Communication and Society,* 15(5), 662–679.

Boyd, D., & Ellison, N. (2007). Social network sites: Definition, history, and scholarship. *Journal of Computer-Mediated Communication,* 13(1), 210–230.

Braman, S. (2006). *Change of state: Information, policy and power.* Cambridge, MA: MIT Press.

Bruns, A. (2008). *Blogs, Wikipedia, Second Life and Beyond: From production to produsage.* New York: Peter Lang.

Bruns, A., & Burgess, J. E. (2011a). New methodologies for researching news discussion on Twitter, The Future of Journalism conference, Cardiff University, Cardiff, UK, September 2011.

Bruns, A., & Burgess, J.E., (2011b). The use of Twitter hashtags in the formation of ad hoc publics, 6th European Consortium for Politics Research conference, University of Iceland, Reykjavik, August 2011.

Bruns, A., & Burgess, J.E., (2012). Notes towards the scientific study of public communication on Twitter. Conference on Science and the Internet, Düsseldorf, Germany, August 2012.

Brunton, F., & Nissenbaum, H. (2011). Vernacular resistance to data collection and analysis: A political theory of obfuscation. *First Monday*, 16(5). Available at: http://firstmonday.org/ojs/index.php/fm/article/view/3493/0. Accessed 21 July 2015.

Bucher, T. (2012). Want to be on the top? Algorithmic power and the threat of invisibility on Facebook. *New Media & Society*, 14(7), 1164–1180.

Bullas, J. (2014). 22 social media facts and statistics you should know in 2014. *Jeffbullas.com*. Available at: http://www.jeffbullas.com/2014/01/17/20-social-media-facts-and-statistics-you-should-know-in-2014/. Accessed 12 Sept 2014.

Burgess, J. (2006). Hearing ordinary voices: Cultural studies, vernacular creativity and digital storytelling. *Continuum: Journal of Media and Cultural Studies*, 20(2), 201–214.

Caraway, B. (2010). Online labour markets: An enquiry into oDesk providers. *Work, Organisation Labour and Globalisation*, 4(2), 111–125.

Carpentier, N., & Dahlgren, P. (eds). (2011). Interrogating audiences: Theoretical horizons of participation. Special issue of *Communication Management Quarterly*, 21. Available at: http://www.cost.eu/media/publications/12-02-Interrogating-audiences-Theoretical-horizons-of-participation-in-CM-Communication-Management-Quarterly. Accessed 20 Mar 2012.

Chandler, D (1998). Personal home pages and the construction of identities on the web. Available at: available http://www.aber.ac.uk/media/Documents/short/webident.html. Accessed 30 Nov 2001.

Coleman, S., Firmstone, J., Kennedy, H., Moss, G., Parry, K., Thornham, H., Thumim, N. (2014). Public engagement and cultures of expertise, RCUK Digital Economy Communities and Cultures Network + Scoping Report. Available at: http://www.communitiesandculture.org/files/2013/01/Scoping-report-Leeds-and-Suggestions.pdf. Accessed 1 Feb 2013.

Collins, K. (2014). Your colleagues pose bigger threat to your privacy than hackers, *Wired*. Available at: http://www.wired.co.uk/news/archive/2014-10/07/privacy-threats-europe-hackers. Accessed 9 Oct 2014.

Coté, M. (2014). Data motility: The materiality of big social data, *Cultural Studies Review*, 20(1). Available at: http://epress.lib.uts.edu.au/journals/index.php/csrj/article/view/3832. Accessed 1 Sept 2014.

Couldry, N. (2014). Inaugural: A necessary disenchantment—Myth, agency and injustice in a digital world. *The Sociological Review*, 62(4), 880–897.

Couldry, N., & Powell, A. (2014). Big data from the bottom up, *Big Data and Society*, 1(1): 1-5. Available at: http://bds.sagepub.com/content/1/2/2053951714539277.full.pdf+html. Accessed 6 Sept 2014.

Crawford, K. (2013). The hidden biases of big data, *Harvard Business Review*, 1 April. Online: http://blogs.hbr.org/cs/2013/04/the_hidden_biases_in_big_data.html. Accessed 6 Jan 2014.

Crawford, K., Lingel, J., & Karppi, T. (2015). Our metrics, ourselves: A hundred years of self-tracking from the weight-scale to the wrist-wearable device.

European Journal of Cultural Studies, 18(405), 479–496.

Cruikshank, B. (2000). *The will to empower: Democratic citizens and other subjects.* Ithaca, NY: Cornell University Press.

D'Orazio, F. (2013). The future of social media research: Or how to re-invent social listening in 10 steps. Retrieved 2 October 2013, from http://abc3d. tumblr.com/post/62887759854/social-data-intelligence. Accessed 6 Nov 2013.

Davies, T., & Edwards, D. (2012). Emerging implications of open and linked data for knowledge sharing in development. *IDS Bulletin, 43*(5), 117–127.

De Ridder, S. (2015). Are digital media institutions shaping youth's intimacy stories? Strategies and tactics in the social networking site Netlog. *New Media and Society, 17*(3), 356–374.

Derrida, J. (1996). *Archive Fever: A Freudian impression,* (trans. E. Prenowitz), Chicago: University of Chicago Press.

Ding, Y., Du, Y., Hu, Y., Liu, Z., Wang, L., Ross, K.W., & Ghose, A. (2011). Broadcast yourself: Understanding YouTube uploaders. Paper presented at the Internet Measurement Conference, IMC'11, 2–4 November, Berlin. Available at: http://conferences.sigcomm.org/imc/2011/program.htm. Accessed 7 May 2014.

Dovey, J. (2006). The revelation of unguessed worlds. In J. Dovey (Ed.), *Fractal dreams: New media in social context.* London: Lawrence and Wishart.

Doyle, G. (2015). Multi-platform media: How newspapers are adapting to the digital era. In K. Oakley & J. O'Connor (Eds.), *Routledge companion to the cultural industries.* London: Routledge.

Driscoll, K., & Walker, S. (2014). Working within a black box? Transparency in the collection and production of big Twitter data. *International Journal of Communication, 8,* 1745–1764.

Duhigg, C. (2009). What does your credit-card company know about you? *The New York Times,* 12th May 2009. Available at: http://www.nytimes. com/2009/05/17/magazine/17credit-t.html?pagewanted=all. Accessed 12 Oct 2012.

Duhigg, C. (2012). How companies learn your secrets. *The New York Times.* Retrieved 16 February 2012, from http://www.nytimes.com/2012/02/19/magazine/shopping-habits.html?pagewanted=all&_r=0. Accessed 12 Oct 2012.

Ellison, N. B., & Boyd, D. (2013). Sociality through social networking sites. In W. H. Dutton (Ed.), *The Oxford handbook of internet studies.* Oxford, England: Oxford University Press.

Espeland, W. N., & Sauder, M. (2007). Rankings and reactivity: How public measures recreate social worlds. *American Journal of Sociology, 113*(1), 1–40.

Facebook (2013). Facebook's Facebook Page. Available at: https://www.facebook.com/facebook.

Facebook Help (n.d.). Ad Basics. Available at: https://www.facebook.com/help/326113794144384. Accessed 27 Apr 2011.

Feenberg, A. (1999). *Questioning technology*. London: Routledge.

Feenberg, A. (2002). *Transforming technology: A critical theory re-visited*. Oxford, England: Oxford University Press.

Foucault, M. (1984) On the genealogy of ethics: An overview of work in progress. In P. Rabinow (Ed), *The Foucault Reader*. New York: Pantheon.

Fowler, G. (2012). When the most personal secrets get outed on Facebook. *The Wall Street Journal*, October 13 2012. Available at: http://online.wsj.com/article/SB10000872396390444165804578008740578200224.html. Accessed 31 Oct 2012.

Fowler, G.A. & de Avila, J. (2009). On the Internet, everyone's a critic but they're not very critical. *Wall Street Journal*, 6 October 2009.

Freire, P. (1970). *Pedagogy of the Oppressed*. New York: Continuum.

Fuchs, C. (2011). Web 2.0, prosumption, and surveillance. *Surveillance and Society*, 8(3), 288–309.

Fuchs, C. (2014). *Social media: A critical introduction*. London: Sage.

Fuchs, C., & Sandoval, M. (2014). *Critique, social media and the information society*. London: Routledge.

Galloway, A. R. (2006). *Gaming: Essays in algorithmic culture*. Minneapolis, MN: University of Minnesota Press.

Gallup, T. N. S. (2012). *Forbrukerrådet—Undersøkelse om digitale tjenetser (Report on digital services)*. Oslo, Norway: The Consumer Council of Norway.

Gehl, R. W. (2014). Power from the C-Suite: The chief knowledge officer and chief learning officer as agents of Noopower. *Communication and Critical/Cultural Studies*, 11(2), 175–194.

Gehl, R. (2015). What is shared and not shared: A partial genealogy of the data scientist. *European Journal of Cultural Studies*, 18(4-5), 379–394.

Gerbaudo, P. (2012). *Tweets and the streets: Social media and contemporary activism*. London: Pluto Books.

Gerlitz, C. (2015). A critique of social media metrics: The production, circulation and performativity of social media metrics. Digital Methods Initiative Winter School, University of Amsterdam, Keynote Speech.

Gerlitz, C., & Lury, C. (2014). Social media and self-evaluating assemblages: On numbers, orderings and values. *Distinktion: Scandinvanian Journal of Social Theory*, 2, 174–188.

Gerlitz, C., & Rieder, B. (2013). Mining One Percent of Twitter: Collections, Baselines, Sampling. *M/C Journal*, 16(2). Available at: http://journal.media-culture.org.au/index.php/mcjournal/article/viewArticle/620. Accessed 27 July 2015.

Gibbs, A. (1997). Focus groups. *Social Research Update*. Available at: http://sru.soc.surrey.ac.uk/SRU19.html. Accessed 20 Dec 2003.

Gibson-Graham, J. K. (2006). *A postcapitalist politics*. Minneapolis, MN: University of Minnesota Press.

Gilbert, J. (2012). Moving on from market society: Culture (and cultural studies) in a post-democratic age. *Open Democracy*. Available at: http://www.opendemocracy.net/ourkingdom/jeremy-gilbert/moving-on-from-market-society-culture-and-cultural-studies-in-post-democra.. Accessed 26 July 2012.

Gill, R. (2007). Technobohemians or the new cybertariat? New media work in Amsterdam a decade after the web. Report for the Institute of Network Cultures. Available at: http://www.lse.ac.uk/collections/genderInstitute/whosWho/profiles/gill.htm. Accessed 26 Sept 2007.

Gill, R. (2010). Life is a pitch: Managing the self in new media work. In M. Deuze (Ed.), *Managing media work*. London: Sage.

Gillespie, T. (2012). The dirty job of keeping Facebook clean. *Culture Digitally*. Retrieved 22 February 2012, from http://culturedigitally.org/2012/02/the-dirty-job-of-keeping-facebook-clean/. Accessed 18 June 2014.

Gillespie, T. (2014). The relevance of algorithms. In T. Gillespie, P. J. Boczkowski, & K. A. Foot (Eds.), *Media technologies: Essays on communication, materiality, and society*. Cambridge, MA: MIT Press. Available at: http://www.tarletongillespie.org/essays/Gillespie%20-%20The%20Relevance%20of%20Algorithms.pdf. Accessed 12 Jan 2015.

Gitelman, L., & Jackson, V. (2013). Introduction. In L. Gitleman (Ed.), *'Raw data' is an oxymoron*. Cambridge, MA: MIT Press.

Gitleman, L. (Ed.). (2013). *Raw data is an oxymoron*. Cambridge, MA: MIT Press.

Gleibs, I. (2014). Turning virtual public spaces into laboratories: Thoughts on conducting online field studies using social network sites. *Analyses of Social Issues and Public Policy*, 14(1), 352–370.

Graham, M., Hale, S. A., & Gaffney, D. (2013). Where in the world are you? Geolocation and language identification in Twitter. *Professional Geographer*. Available at: http://arxiv.org/ftp/arxiv/papers/1308/1308.0683.pdf. Accessed 12th May 2014.

Graham, T., & Wright, S. (2014). Discursive equality and everyday talk online: The impact of "superparticipants". *Journal of Computer-Mediated Communication*, 19, 625–642.

Granovetter, M. (1985). Economic action and social structure: The problem of embeddedness. *American Journal of Sociology*, 91(3), 481–510.

Gregg, M. (2011). *Work's intimacy*. Stafford, BC, Australia: Polity Press.

Gregg, M. (2015). Hack for good: Speculative labor, app development and the burden of austerity. Available at: http://www.academia.edu/8728551/Hack_for_good_Speculative_labor_app_development_and_the_burden_of_austerity. Accessed 28 Apr 2015.

Grosser, B. (2014). What do metrics want? How quantification prescribes social interaction on Facebook. *Computational Culture: A Journal of Software Studies*, 4.

Available at: http://computationalculture.net/article/what-do-metrics-want. Accessed 10 Jan 2015.

Gurstein, M. B. (2011). Open data: Empowering the empowered or effective data use for everyone? *First Monday*, 16, 2–7. Available at: http://journals.uic.edu/ojs/index.php/fm/article/view/3316/2764. Accessed 28th Apr 2015.

Hammersley, M. (2002). Action research: A contradiction in terms? British Educational Research Association. Exeter, England, 12-14 September 2002. Available at: http://www.leeds.ac.uk/educol/documents/00002130.htm. Accessed 12 May 2014.

Haraway, D. (1985). A manifesto for cyborgs: Science, technology and socialist-feminism in the 1980s. *Socialist Review*, 80, 65–107.

Harper, D., Tucker, I., & Ellis, D. (2014). Surveillance and subjectivity: Everyday experiences of surveillance. In K. Ball & L. Snider (Eds.), *The surveillance-industrial complex: A political economy of surveillance*. London: Routledge.

Harvey, P., Reeves, M., & Ruppert, E. (2013). Anticipating failure: Transparency devices and their effects. *Journal of Cultural Economy*, 3(6), 294–312.

Havalais, A. (2013). Home made big data? Challenges and opportunities for participatory social research. *First Monday*. 10 Feb 2014. Available at: http://firstmonday.org/ojs/index.php/fm/article/view/4876.. Accessed 11 Feb 2014.

Hayward, C. (2012). We Know What You're Doing website. Available at: http://www.weknowwhatyouredoing.com/. Accessed 4 June 2012.

Hearn, A. (2008). "Meat, mask, burden": Probing the contours of the branded "self". *Journal of Consumer Culture*, 8(2), 197–217.

Hearn, A. (2010). Structuring feeling: Web 2.0, online ranking and rating, and the digital "reputation" economy. *Ephemera: Theory & Politics in Organisation*, 10(3/4). Available at: http://www.ephemeraweb.org/. Accessed 11 Feb 2011.

Hearn, A. (2013). Tracing worker subjectivities in the data stream. Paper at International Communication Association (ICA) Annual Conference, London, 2013.

Helles, R., & Bruhn Jensen, K. (2013), 'Introduction' to special issue Making data—Big data and beyond. *First Monday*, 10 Feb 2014. Available at: http://firstmonday.org/ojs/index.php/fm/issue/view/404. Accessed 11 Feb 2014.

Helmond, A., & Gerlitz, C. (2013). The like economy: Social buttons and the data-intensive web. *New Media and Society*, 15(8), 1348–1365.

Hern, A. (2014). Sir Tim Berners-Lee speaks out on data ownership. *The Guardian*. Available at: http://www.theguardian.com/technology/2014/oct/08/sir-tim-berners-lee-speaks-out-on-data-ownership. Accessed 9 Oct 2014.

Hesmondhalgh, D. (2010). Russell Keat, cultural goods and the limits of the market. *International Journal of Cultural Policy*, 1615(1), 37–38.

Hesmondhalgh, D. and Baker, S. (2010) *Creative Labour: media work in three cultural industries*, London: Routledge.

Hesmondhalgh, D. (2013). *Why music matters*. Chichester, England: Wiley Blackwell.

Hesmondhalgh, D. (2014). Applying the capabilities approach to cultural production. *International Communication Association International Conference*, May 2014, Seattle.

Hill, K. (2012). How Target figured out a teen girl was pregnant before her father did. *Forbes*. Retrieved 16 February 2012, from http://Www.Forbes.Com/ Sites/Kashmirhill/2012/02/16/How-Target-Figured-Out-A-Teen-Girl-Was-Pregnant-Before-Her-Father-Did/. Accessed 12 Oct 2012.

Hintz (2014). Policy hacking: Hackathons and policy code. *Communication for Empowerment: Citizens, Markets, Innovations*, ECREA Conference, Lisbon, Portugal, 14 November 2014.

Hofer-Shall, Z., Stanhope, J., & Smith, A. (2012). The social intelligence market is immature. Forrester Report.

Horkheimer, M., & Adorno, T. (1948). *Dialectic of enlightenment*. Frankfurt, Germany: Fischer.

Howard, A. (2011). Making open government data visualizations that matter. govfresh, Retrieved 13 March 2011, from http://gov20.govfresh.com/making-open-government-data-visualizations-that-matter/. Accessed 27 Apr 2015.

Huff, D. (1954). *How to lie with statistics*. London: Penguin.

Humphreys, L. (2011). Who's watching whom? A study of interactive technology and surveillance. *Journal of Communication*, 61, 575–595.

Hunt, T. (2009). *The Whuffie factor: Using the power of social networks to build your business*. New York: Crown Business.

Irani, L. (2015a). Difference and dependence among digital workers: The case of Amazon Mechanical Turk. *South Atlantic Quarterly*, 114(1).

Irani, L. (2015b). Hackathons and the making of entrepreneurial citizenship. *Science, Technology and Human Values*. Available at: http://sth.sagepub.com/content/early/2015/04/07/0162243915578486.abstract. Accessed 23rd July 2014.

Irani, L., & Silberman, M.S. (2013). Turkopticon: Interrupting worker invisibility in Amazon Mechanical Turk. *Proceedings of CHI 2013*, April 28th–May 2nd, 2013.

Ito, M., Baumer, S., Bittani, M., Boyd, D., Cody, R., Herr-Stephenson, B., et al. (2009). *Hanging out, messing around and geeking out: Kids living and learning with new media*. Cambridge, MA: MIT Press.

Jayson, S. (2014). Social media research raises privacy and ethics issues. *USA Today*. Retrieved 12 March 2014, from http://www.usatoday.com/story/news/nation/2014/03/08/data-online-behavior-research/5781447/. Accessed 28 Apr 2015.

Jenkins, H. (2008). *Convergence culture: Where new and old media collide*. New York: New York University Press.

John, N. (2013). Sharing and web 2.0: The emergence of a keyword. *New Media and Society*, 15(2), 167–182.

Johnson, B. (2010). Privacy no longer a social norm, says Facebook founder. *The Guardian* website. Available at: http://www.theguardian.com/technology/2010/jan/11/facebook-privacy. Accessed 7th May 2011.

Kaplan, A. M., & Haenlein, M. (2010). Users of the world, unite! The challenges and opportunities of social media. *Business Horizons,* 53(1), 59–68.

Kawash, J. (2015). *Online social media analysis and visualisation (Lecture notes in social networks).* New York: Springer.

Keat, R. (2000). Every economy is a moral economy. Available at: http://www.russellkeat.net/research/ethicsmarkets/keat_everyeconomymoraleconomy.pdf.. Accessed 26 July 2012.

Keat, R. (2011). Market economies as moral economies. Available at: http://www.russellkeat.net/research/ethicsmarkets/keat_marketeconomies_moraleconomies.pdf. Accessed 26 July 2012.

Kember, S., & Zylinska, J. (2012). *Life after new media: Mediation as a vital process.* Cambridge, MA: MIT Press.

Kennedy, H. (2006). Beyond anonymity, or future directions for Internet identity research. *New Media and Society,* 8(6), 859–876.

Kennedy, H. (2011). *Net work: Ethics and values in web design.* Basingstoke, England: Palgrave MacMillan.

Kennedy, H., Elgesem, D., & Miguel, C. (2015). On fairness: User perspectives on social media data mining. *Convergence* 10.1177/1354856515592507.

Kennedy, H. (2015) 'Seeing Data', LSE Impact Blog. Retrieved July 22, 2015, from http://blogs.lse.ac.uk/impactofsocialsciences/2015/07/22/seeing-data-how-people-engage-with-data-visualisations/

Kennedy, H., & Moss, G. (2015). Known or knowing publics? Social media data mining and the question of public agency. *Big Data and Society* (forthcoming).

Kennedy, H., Moss, G., Birchall, C., & Moshonas, S. (2015). Balancing the potential and problems of digital data through action research: Methodological reflections. *Information Communication and Society,* 1892, 172–186.

Krueger, R., & Casey, M. A. (2008). *Focus groups: A practical guide for applied research.* London: Sage.

Kwak, H., Lee, C., Park, H., & Moon, S. (2010). What is Twitter, a social network or a news media? *Proceedings of the 19th International World Wide Web (WWW) Conference,* April 26–30, Raleigh NC, 591–600. Available at http://an.kaist.ac.kr/traces/WWW2010.html l. Accessed 28 Apr 2014.

Law, J. (1987). Technology and heterogenous engineering: The case of Portuguese expansion. In W. Bijker, T.P. Hughes, & T. Pinch (Eds.), *The Social Construction of Technological Systems.* Cambridge, MA: MIT Press.

Lash, S. (2007). Power after hegemony: Cultural studies in mutation. *Theory Culture & Society,* 24(3), 55–78.

Law, J., Ruppert, E., & Savage, M. (2011). The double social life of methods. *CRESC Working Paper,* 95. Available at: http://www.open.ac.uk/research-projects/iccm/library/164. Accessed 4 June 2012.

Law, J., & Urry, J. (2004). Enacting the social. *Economy and Society*, 33(3), 390–410.

Layder, D. (2006). *Understanding social theory* (2nd ed.). London: Sage.

Lister, M., Dovey, J., Giddens, S., Grant, I., & Kelly, K. (2009). *New media: A critical introduction* (2nd ed.). London: Routledge.

Lorenzen, S. (2015). Keynote, Digital Methods Initiative Winter School, University of Amsterdam.

Lowndes, V., & Squires, S. (2012). Cuts, collaboration and creativity. *Public Money & Management*, 32, 401–408.

Lusoli, W., Bacigalupo, M., Lupianez, F., Andrade, N., Monteleone, S., & Maghiros, I. (2012). *Pan-European Survey of Practices, Attitudes and Policy Preferences as regards Personal Identity Data Management*. Available at: http://is.jrc.ec.europa.eu/pages/TFS/documents/EIDSURVEY_Web_001.pdf. Accessed 19 June 2013.

Lyon, D. (2014). Surveillance, Snowden, and big data: Capacities, consequences, critique. *Big Data and Society*, 1(1), 1–12. Available at: http://bds.sagepub.com/content/1/2/2053951714541861.full.pdf+html. Accessed 10 Oct 2014.

Lyon, D., & Bauman, Z. (2012). *Liquid surveillance: A conversation*. Cambridge, MA: Polity Press.

Mackenzie, A. (2013). Programming subjects in the regime of anticipation: Software studies and subjectivity. *Subjectivity*, 6, 391–405.

Madden, M. (2012). Privacy management on social media sites. Available at: http://www.pewinternet.org/2012/02/24/privacy-management-on-social-media-sites/. Accessed 20 July 2013.

Madden, M., & Smith, A. (2010). Reputation management and social media: How people monitor their identity and search for others online. Available at: http://pewinternet.org/~/media/Files/Reports/2010/PIP_Reputation_Management.pdf. Accessed 20 July 2013.

Manovich, L. (2011). Trending: The promises and the challenges of big social data. Available at: http://www.manovich.net/DOCS/Manovich_trending_paper.pdf. (Also in M. K. Gold (Ed.), *Debates in the Digital Humanities*). Accessed 9 Oct 2013.

Marotta-Wurgler, F. (2014). Talk given to NYU MCC/Law Privacy Research Group. Retrieved 9 April 2014, from Profile available at: https://its.law.nyu.edu/facultyprofiles/profile.cfm?section=pubs&personID=27875.

Marres, N. (2012). The redistribution of methods: On intervention in digital social research, broadly conceived. *Live Methods: Sociological Review Monographs*, 60, 139-165.

Marres, N. (2015). Are we mapping society or technology? Analysing "privacy" issues with Twitter. University of Sheffield. Retrieved 23 April 2015, from http://blogs.cim.warwick.ac.uk/complexity/wp-content/uploads/sites/11/2014/11/Marres_society_technology.pdf. Accessed 20 July 2015.

Marres, N., & Gerlitz, C. (2015). Interface methods: Renegotiating relations between digital social research, STS and sociology. *Sociological Review* (forthcoming).

Martin, K. (2012). Diminished or just different? A factorial vignette study of privacy as a social contract. *Journal of Business Ethics* 111(4): PAGE. Online, available at http://papers.ssrn.com/sol3/papers.cfm?abstract_id=2280730.

Marwick, A. (2012). The public domain: Social surveillance in everyday life. *Surveillance & Society*, 9(4), 378–393.

Marwick, A., & Boyd, D. (2010). I tweet honestly, I tweet passionately: Twitter users, context collapse, and the imagined audience. *New Media and Society*, 13, 96–113.

Marwick, A., & Boyd, D. (2014). Networked privacy: How teenagers negotiate context in social media. *New Media and Society*, 16, 1051–1967.

Marx, K. (1852). *The Eighteenth Brumaire of Louis Napoleon*. Available at: https://www.marxists.org/archive/marx/works/1852/18th-brumaire/ch01.htm. Accessed 21 Dec 2012.

Massumi, B. (2002). *Parables of the virtual: movement, affect, sensation*. Durham, NC: Duke University Press.

Mayer-Schoenberger, V., & Cukier, K. (2013). *Big data: A revolution that will transform how we live, work and think*. London: John Murray.

McCarthy, A. (2008). From the ordinary to the concrete: Cultural studies and the politics of scale. In M. White & J. Schwoch (Eds.), *Questions of method in cultural studies*. Oxford, England: Wiley Blackwell.

McCarthy, T. (2012). Mark Zuckerberg's sister learns life lesson after Facebook photo flap. *The Guardian*, 27 December 2012. Available at: http://www.theguardian.com/technology/us-news-blog/2012/dec/27/facebook-founder-sister-zuckerberg-photo. Accessed 12 Mar 2013.

McKelvey, F., Tiessen, M., & Simcoe, L. (2015). A consensual hallucination no more? The Internet as simulation machine. *European Journal of Cultural Studies*, 18(4–5), 577–594.

McQuillan, D. (2015). Algorithmic states of exception. *European Journal of Cultural Studies*, 18(4–5), 564–576.

McStay, A. (2009). *Digital advertising*. Basingstoke, England: Palgrave MacMillan.

McStay, A. (2011). *The mood of information: A critique of online behavioural advertising*. London: Continuum International Publishing Group.

Milan, S. (2014). Data activism: The politics of "big data" in the throes of civil society. *Communication for Empowerment: Citizens, Markets, Innovations*, ECREA Conference, Lisbon, Portugal, 13 Nov 2014.

Mindruta, R. (2013). Top 10 free social media monitoring tools. Brandwatch blogpost. Retrieved 9 August 2013, from http://www.brandwatch.com/2013/08/top-10-free-social-media-monitoring-tools/. Accessed 21 Apr 2015.

Moss, G., Kennedy, H., Moshonas, S., & Birchall, C. (2015). Knowing your publics: The use of social media analytics in local government. *Journal of Information Technology and Polity*, (forthcoming).

Mukerji, C., & Schudson, M. (Eds.). (1991). *Rethinking popular culture*. Berkeley, CA: University of California Press.

Munson, S. (2014). Building and breaking bubbles: Designing technological systems that select and present news. *Communication for Empowerment: Citizens, Markets, Innovations*, ECREA Conference, Lisbon, Portugal, 13 Nov 2014.

Murakami Wood, D., Ball, K., Lyon, D., Norris, C., & Raab, C. (2006). A Report on the Surveillance Society. Report for the UK Information Commissioner's Office. *Surveillance Studies Network, UK*. Available at: http://ico.org.uk/~/media/documents/library/Data_Protection/Practical_application/SURVEILLANCE_SOCIETY_FUL L_REPORT_2006.ashx. Accessed 29 Apr 2015.

Nafus, D., & Sherman, J. (2014). This one does not go up to 11: The quantified self movement as an alternative big data practice. *International Journal of Communication, 8*, 1784–1794.

Nesbitt, H. (2012). Deterritorial support group. *Dazed*. Available at: http://www.dazeddigital.com/artsandculture/article/10260/1/deterritorial-support-group. Accessed 27 July 2015.

Netichailova, E. (2012). Facebook as a surveillance tool: From the perspective of the user. *Communication, Capitalism & Critique, 10*(2), 683–691.

Neurath, O. (1973). Empiricism and sociology. In M. Neurath & R. S. Cohen (Eds.), *Vienna circle collection*. Dordrecht, The Netherlands: Reidel.

Nissenbaum, H. (2009). *Privacy in context: Technology, policy and the integrity of social life*. Stanford, CA: Stanford University Press.

O'Connor, M. (2013). 10 top social media monitoring and analytics tools. TweakYourBiz post. Retrieved 6 March 2013, from http://tweakyourbiz.com/marketing/2013/03/06/10-top-social-media-monitoring-analytics-tools/. Accessed 21 Apr 2015.

O'Neill, J. (1998). *The market: Ethics, knowledge and politics*. London: Routledge.

O'Reilly, T. (2005). What is Web 2.0? Design patterns and business models for the next generation of software. O'Reilly website. Available at: http://www.oreillynet.com/lpt/a/6228. Accessed 4 Aug 2010.

ONS (2013). Internet access: Households and individuals, 2013. ONS website. Available at: http://www.ons.gov.uk/ons/dcp171778_322713.pdf. Accessed 12 Sept 2014.

Oxford Internet Institute (n.d.). Accessing and using big data to advance social scientific knowledge. Available at: http://www.oii.ox.ac.uk/research/projects/?id=98. Accessed 12 Dec 2013.

Papacharissi, Z. (Ed.). (2010). *A private sphere: Democracy in a digital age*. Cambridge, MA: Polity Press.

Perlin, R. (2011). *Intern nation: How to earn nothing and learn little in the brave new economy*. London: Verso Books.

Peters, J. D. (1995). Historical tensions in the concept of public opinion. In T. L. Glasser & C. T. Salmon (Eds.), *Public opinion and the communication of consent*. New York: Guilford Press.

Pinch, T. (1992). Opening black boxes: Science, technology and society. *Social Studies of Science*, 22(3), 487–510.

Pinch, T., & Bijker, W. (1987). The social construction of facts and artefacts. In W. Bijker, T. Hughes, & T. Pinch (Eds.), *The social construction of technological systems*. Cambridge, MA: MIT Press.

Polsky, N. (1971). *Hustlers, beats, and others*. Pscataway, NJ: Aldine.

Porter, T. M. (1995). *Trust in numbers: the pursuit of objectivity in science and public life.*. Princeton: Princeton University Press.

Porter, T. M. (1996). *Trust in numbers: The pursuit of objectivity in science and public life*. Princeton, NJ: Princeton University Press.

Pybus, J., Coté, M., Blanke, T. (2015). Hacking the social life of big data: A data literacy framework, *Big Data and Society*, 10.1177/2053951715616649.

Rae, A. (2014). Open data visualization: The dawn of understanding?, StatsLife Blog, Royal Statistical Society. Retrieved 16 September 2014, from http://www.statslife.org.uk/opinion/1815-open-data-visualisation-the-dawn-of-understanding. Accessed 29 Sept 2014.

Ravetz, J.R. (1996). In numbers we trust. *Issues in Science and Technology*. Available at: http://www.issues.org/13.2/ravetz.htm. Accessed 12 July 2015.

Raynes-Goldie, K. (2010). Aliases, creeping, and wall cleaning: Understanding privacy in the age of Facebook. *First Monday* 15(1). Available at: http://firstmonday.org/article/view/2775/2432. Accessed 19 June 2013.

Reason, P., & Bradbury, H. (2001). Introduction. In P. Reason & H. Bradbury (Eds.), *Handbook of action research: Participative inquiry and practice*. London: Sage.

Rogers, R. (2013). *Digital methods*. Cambridge, MA: The MIT Press.

Roginsky, S. (2014). Research Seminar, Institute of Communications Studies, University of Leeds, 6 May 2014.

Ross, A. (2003). *No-Collar: The humane workplace and its hidden costs*. Philadelphia: Temple University Press.

Ruppert, E., Law, J., & Savage, M. (2013). Reassembling social science methods: The challenge of digital devices. *Theory Culture and Society*, 30(4), 22–46.

Russell, M.A. (2013). *Mining the Social Web: Data mining Facebook, Twitter, LinkedIn, google+, GitHub and more*, O'Reilly Media

Sayer, A. (2004). Moral economy, Department of Sociology, Lancaster University. Available at: www.lancs.ac.uk/fass/sociology/papers/sayer-moral-economy.pdf. Accessed 12 Sept 2010.

Sayer, A. (2011). *Why things matter to people: Social science, values and ethical life*. Cambridge, MA: Cambridge University Press.

Serres, M. (2007). *The parasite*. Minneapolis, MN: University of Minnesota Press.

Shenk, D. (1997). *Data smog*. New York: Harper Collins.

Silverstone, R. (1994). The power of the ordinary: On cultural studies and the sociology of culture. *Sociology*, 28(4), 991–1001.

Sledge, M. (2013). CIA's Gus Hunt On Big Data: We "Try To Collect Everything And Hang On To It Forever." *Huffington Post*, March 20. Available at: http://www.huffingtonpost.com/2013/03/20/cia-gus-hunt-big-data_n_2917842.html. Accessed 28 Apr 2015.

Smith, A. (2014). The 11 providers that matter most and how they stack up. Forrester Wave TM Enterprise Listening Report.

Social Media Biz (2011). Top 20 social media monitoring vendors for business. Retrieved 12 January 2011, from http://socialmedia.biz/2011/01/12/top-20-social-media-monitoring-vendors-for-business/. Accessed 31 Oct 2013.

Social Media Today (2013). 50 top tools for social media monitoring, analytics and management. Retrieved 16 May 2013, from http://socialmediatoday.com/pamdyer/1458746/50-top-tools-social-media-monitoring-analytics-and-management-2013. Accessed 31 Oct 2013.

Stengers, I., & Zournazi, M. (2003). A "Cosmo-Politics"—Risk, hope, change. In M. Zournazi (Ed.), *Hope: New philosophies for change*. New York: Routledge.

Strathern, M. (2000). Introduction: New accountabilities. In M. Strathern (Ed.), *Audit cultures: Anthropological studies in accountability, ethics and the academy*. London: Routledge.

Striphas, T. (2015). Algorithmic culture. *European Journal of Cultural Studies, 18*(4–5), 395–412.

Sysomos (2012). Sysomos website. Available at: http://www.sysomos.com/solutions/. Accessed 31 July 2012.

Szabo, G., & Boykin, O. (2015). *Social Media Data Mining and Analytics*, John Wiley Media.

Thelwall, M., Buckley, K., & Paltoglou, G. (2011). Sentiment in Twitter events. *Journal of The American Society for Information Science and Technology, 62*(2), 406–418.

Thumim, N. (2012). *Self-representation and digital culture*. Basingstoke, England: Palgrave MacMillan.

Tkacz, N. (2014). Connection perfected: What the dashboard reveals. Digital Methods Initiative Winter School, University of Amsterdam, Keynote Speech.

Totka, M. (2014). Top 10 monitoring tools for Twitter and other social media platforms. Social Media Biz post. Retrieved 12 February 2014, from http://socialmedia.biz/2014/02/12/top-10-monitoring-tools-for-twitter-other-social-media-platforms/. Accessed 21 Apr 2015.

Toynbee, J. (2007). *Bob Marley: Herald of a postcolonial world*. Cambridge, MA: Polity Press.

Trepte, S., & Reinecke, L. (2011). The social web as a shelter for privacy and authentic living. In S. Trepte & L. Reinecke (Eds.), *Privacy online: Perspectives on privacy and self-disclosure on the social web*. New York: Springer.

Trottier, D. (2012). *Social media as surveillance: Rethinking visibility in a converging world*. Farnham, England: Ashgate Press.

Turow, J. (1997). *Breaking up America: Advertisers and the new media world.* Chicago: University of Chicago Press.

Turow, J. (2008). *Niche Envy: Marketing discrimination in the digital age.* Boston: MIT Press.

Turow, J. (2012). *The Daily You: How the new advertising industry is defining your identity and your worth.* New Haven, CT: Yale University Press.

Turow J, Feldman L and Meltzer K (2005). Open To Exploitation: American shoppers, online and offline. Available at: http://www.annenbergpublicpolicy-center.org/Downloads/Information_And_Society/Turow_APPC_Report_WEB_FINAL.pdf. Accessed 25 July 2012.

Turow, J., Hennessey, M., & Draper, N. (2015a). *The Trade Off Fallacy: How marketers are misrepresenting American consumers and opening them up to exploitation.* Available at: https://www.asc.upenn.edu/sites/default/files/TradeoffFallacy_1.pdf. Accessed 12th June 2015.

Turow J, Hoofnagle CJ, King J, Bleakley A and Hennessy M (2009). Americans Reject Tailored Advertising and Three Activities that Enable it. Available at: http://www.ftc.gov/bcp/workshops/privacyroundtables/Turow.pdf. Accessed 25 July 2012.

Turow, J., McGuigan, L., & Maris, E. R. (2015b). Making data mining a natural part of life: Physical retailing, customer surveillance and the 21st century social imaginary. *European Journal of Cultural Studies,* 18(4–5), 464–478.

Twitter (2013). Celebrating #Twitter7. 21 March 2013. Retrieved from, https://blog.twitter.com/2013/celebrating-twitter7.

Twitter (2014). Twitter.com – Privacy Policy. Available at: www.docracy.com/pdf/0pes0jdczo3/1. Accessed 18 March 2015.

Twitter Analytics (n.d.). Twitter Analytics Website, analytics.twitter.com. Accessed 6 Sept 2014.

van Dijck, J. (2009). Users like you? Theorising agency in user-generated content. *Media Culture and Society,* 31(1), 41–58.

van Dijck, J. (2013a). *The culture of connectivity: A critical history of social media.* Oxford, England: Oxford University Press.

van Dijck, J. (2013b). You have one identity: Performing the self on Facebook and LinkedIn. *Media, Culture & Society,* 35(2), 199–215.

van Dijck, J. (2014). Datafication, dataism and dataveillance: Big data between scientific paradigm and ideology. *Surveillance and Society,* 12(2), 197–208.

van Dijck, J., & Poell, T. (2013). Understanding social media logic. *Media and Communication,* 1(1), 2–14.

van Zoonen, L. (2001). Desire and resistance: Big Brother and the recognition of everyday life. *Media Culture and Society,* 23(5), 667–9.

van Zoonen, L. (2014). The data delirium. https://liesbetvanzoonen.wordpress.com/page/3/.

Vis, F. (2013). A critical reflection on Big Data: Considering APIs, researchers and tools as data makers. *First Monday*, 18(10). Available at: http://firstmonday. org/ojs/index.php/fm/article/view/4878. Accessed 11 Feb 2014.

Vis, F., & Thelwall, M. (2016, forthcoming) *Researching social media.*, London: Sage Publications.

Webopedia (2014). How much data is out there? Webopedia. Retrieved 6 January 2014, from http://www.webopedia.com/quick_ref/just-how-much-data-is-out-there.html. Accessed 9 June 2014.

Webster, J. (1999). *Shaping women's work: Gender, employment and information technology*. London: Routledge.

Weller, K., Bruns, A., Burgess, J., Mahrt, M., & Puschmann, C. (2013). Twitter and society: An introduction. In K. Weller, A. Bruns, J. Burgess, M. Mahrt, & C. Puschmann (Eds.), *Twitter and society*. New York: Peter Lang.

Williams, R. (2013). Culture is ordinary. In J. McGuigan (Ed.), *Raymond Williams on culture and society: Essential writings*. London: Sage. Original work published on 1958.

Williams, R. (1961). *The long revolution*. London: Chatto and Windus.

Williamson, B. (2014). The death of the theorist and the emergence of data and algorithms in digital social research. *Impact of Social Sciences*, LSE Blog. Retrieved 10 February 2014, from http://blogs.lse.ac.uk/impactofsocialsciences/2014/02/10/the-death-of-the-theorist-in-digital-social-research/. Accessed 11 Feb 2014.

Wittel, A. (2011). Qualities of sharing and their transformations in the digital age. *International Review of Information Ethics*, 15, 3–8.

Wood, D. (2005). Editorial: People watching people. *Surveillance and Society*, 2, 474–478.

Wyatt, S. (1998). *Technology's arrow: Developing information networks for public administration in Britain and the United States*. Maastricht, The Netherlands: University of Maastricht.

Young, A., & Quan-Haase, A. (2013). Privacy protection strategies on Facebook. *Information Communication and Society*, 16(4), 479–500.

Zada, J. (2011). Take This Lollipop. Available at http://www.takethislollipop.com/. Accessed 4 Dec 2011.

Zafarani, R., Ali Abbasi, A., & Liu, H. (2014). *Social media mining: An introduction*. Cambridge, MA: Cambridge University Press.

Zambrano, R.N., & Engelhardt, Y. (2008). Diagrams for the masses: Raising public awareness – from Neurath to Gapminder and Google Earth. *Diagrammatic Representation and Inference* (Lecture Notes in Computer Science) 5223: 282–292, New York: Springer. Available at: 10.1007/978-3-540-87730-1_26.

INDEX[1]

[1] Note: Page number followed by 'n' refers to footnotes.

© The Editor(s) (if applicable) and The Author(s) 2016
H. Kennedy, *Post, Mine, Repeat*,
DOI 10.1057/978-1-137-35398-6

—

Printed by Printforce, the Netherlands